REARING GAME BIRDS
AND GAMEKEEPING

REARING GAME BIRDS
AND GAMEKEEPING

Management
techniques for
pheasant and
partridge

BETH WILLIAMS

Copyright © 2013 Beth Williams

First published in the UK in 2013
by Quiller, an imprint of Quiller Publishing Ltd

British Library Cataloguing-in-Publication Data
A catalogue record for this book is available from the British Library

ISBN 978 1 84689 144 1

The right of Beth Williams to be identified as the author of this work has been asserted in accordance with the Copyright, Design and Patent Act 1988.

The information in this book is true and complete to the best of our knowledge. All recommendations are made without any guarantee on the part of the Publisher, who also disclaims any liability incurred in connection with the use of this data or specific details.

All rights reserved. No part of this book may be reproduced or transmitted in any form or by any means, electronic or mechanical including photocopying, recording or by any information storage and retrieval system, without permission from the Publisher in writing.

All photographs and images by, or property of, the author, except where stated.

Printed in China

Book design by Sharyn Troughton

Illustrations by Eilidh Muldoon

Quiller
An imprint of Quiller Publishing Ltd
Wykey House, Wykey, Shrewsbury, SY4 1JA
Tel: 01939 261616 Fax: 01939 261606
E-mail: info@quillerbooks.com
Website: www.countrybooksdirect.com

CONTENTS

ACKNOWLEDGEMENTS	8
INTRODUCTION	9
Our Beginnings	10
Start Right	12
1. BACKGROUND AND BUYING-IN THE STOCK	15
Putting Pheasants and Partridges into Context	15
Pheasant History	16
The 'Modern' Pheasant; Less Common Varieties of Pheasant	
The Common Pheasant *Phasianus colchicus*	23
Characteristics	
The Partridge Family	26
The Grey Partridge; Other Family Members; The Red-leg (French) Partridge	
Getting Started with Rearing	31
Sourcing Stock – General Guidelines; Four Options; Delivery; Our Initial Experience	
2. EGG-LAYING AND INCUBATION	42
The Egg	42
Laying	44
Natural versus Artificial Incubation; Laying Cycles; Raised Pens	
Incubators	46
Number of Incubators; Temperature and Humidity Measuring Equipment; Buying Your Machine	

The Incubation Process   52
    Gathering and Setting the Eggs; Storing Eggs; Setting the Eggs; Candling;
    The Hatch; Assisted Hatching; Premature or Overdue Chicks
Alternative Use of the Incubating Machinery   67

3. THE FIRST GROWTH PERIOD   69
The Incubation Suite   70
    The Post-natal Nursery; Malformations; The Nursery Base;
    Introducing Food; Preparation and Routine
The Brooders   76
    Construction; Equipment; Early Feeding and Health Issues
The Runs   83
    Bits and De-beaking
The Rearing Pens   85
    Construction; Interior Furnishings; General Routine

4. RELEASE PENS   93
Release Pen Principles   93
    The 'Gentle Release' System; The Open-top Pheasant Release Pen;
    The Site; Construction; Introducing the Birds; Wing-clipping;
    A Note on 'Batches'; The Transfer; Husbandry Post-transfer
The Partridge Release Process   113
    Partridge Release Pen Construction; Introducing the Birds
'Dogging-in'   118

5. BROODY HENS AND STOCK MAINTENANCE   120
Broody Hens   120
    The Broody Breeds; Hen Coops; Brooding
Meanwhile…   126
The Nesting Game Bird   127
    Pheasants; Partridges
Handling the Surplus   131

6. HYGIENE AND MANAGING DISEASE   134
Interim Clean-up   134
Disease   135
    Early Experiences; Proactive Practices; Common Diseases

7. GAME CROPS AND LEGAL COVER   144
Analysis and Preparation of the Land   145
Climate   146

What Crops When and How?	147
Tried and Tested; Perennials; Mix and Match; Catch Crops; Organic or Not?	
Getting Prepared	154
Rotation, Weedbed Control and Seedbed Preparation; Pests; Trouble-shooting Dos and Don'ts	
Costs and Conservation	161
Why Take Risks?	162
Are Your Legal Arrangements in Order?	167
8. PEST AND PREDATOR CONTROL	172
Rodent Control	173
Poisoning Mammalian Pests	173
Conventional Predators	174
Scientific Background to Predator Control; The Law and Licences; Predator Behaviour and Control Options	
Firearms and Their Use	183
Firearms Law in the UK; Technical Issues and Choice of Firearms; Safety; Equipment Preparation and Practice; The Humane Kill; Shooting at Night	
Trapping and Snaring Mammals	194
Trapping and the Law; Using Traps for Mammals; Snares	
Trapping Pest Birds	201
Legal Issues; Traps; Points to Remember	
Less Conventional Deterrents	205
Mechanical/Artificial Devices; Natural Deterrents; Mix and Match	
9. A FINAL FEW WORDS	210
Review of Our Objectives	210
Global Reminders	211
The Professionals' Postscript	214
The British Association of Shooting & Conservation (BASC); The Countryside Alliance Foundation (CAF); Game Farmers' Association (GFA); Game & Wildlife Conservation Trust (GWCT); National Gamekeepers' Organisation (NGO); The Keepers Themselves	
REFERENCES AND RESOURCES	218
INDEX	221

ACKNOWLEDGEMENTS

First, heartfelt thanks to my husband who, not known for his patience, has shown remarkable dedication to his newfound chick midwifery skills; whilst handling the technical side of the shoot with his usual levels of competence. Sincere thanks also to my sister Di, to Lisa, Tony and Trish who have been unstinting with their support. And, of course, to Eilidh Muldoon for her superb illustrations which feature throughout the book. I'm also grateful to all who supplied photographs and other images for use in this book – their kindness has been acknowledged on the appropriate pages.

I am further indebted to the following people. They are all specialists, and each has been extraordinarily generous with their support and advice.

- **Andrew Johnston.** Managing Director, Quiller Publishing Ltd.
- **Bill MacFarlane.** Game farmer, MacFarlane Pheasants.
- **Charles Nodder.** Writer, and adviser to the GFA and NGO.
- **David Taylor.** Shooting Campaign Manager, CAF.
- **Dr Joanne Cooper.** Curator of the modern bird skeleton collection, Natural History Museum.
- **Dr Mike Swan.** Head of Education, GWCT.
- **Glynn Evans.** Head of Gamekeeping, BASC.
- **Jeffrey Olstead.** Head of Publications, BASC.
- **Martin Titley.** Marketing Director, Forage and Amenity Seeds, Limagrain UK Ltd.
- **Nick Pardoe.** Game Farmer, Bonson Wood Game Farm.
- **Richard Barnes.** Game Cover and Conservation Crops Manager, Kings Game Cover.
- **Richard Byas.** BVetMed, MRCVS, Sandhill Veterinary Services.
- **Richard Crofts.** Stock and Shooting Sales Manager, Bettws Hall Game Farm.
- **Tom Devey.** Rural affairs legal expert, MFG Solicitors LLP.
- **Zoe Hunter.** Game farmer, Allandoo Pheasantry.

And finally, all those gamekeepers too numerous to mention, but they know who they are – thank you all!

INTRODUCTION

'WHY?'

This book describes our involvement in raising feathered game from scratch to release, starting off with a couple of converted dog kennels, an outhouse, and some land. It's designed to be a helpful guide, supplying enough information to allow you the flexibility to do things the way you wish.

When we set out on this project we made several mistakes, none of which we're proud of. But we learnt through experience, and now maintain a robust rearing programme which achieves our original aims successfully. To help you avoid making the same errors, I've provided advice from experts in each subject area which will give you an early advantage.

Throughout, the techniques we have used are all geared towards rearing pheasants and red-leg partridges in relatively small batches. That said, the fundamental principles are equally applicable to most of the commonly shot game birds in Britain.

Our annual target number is release to wood of between 250 and 350 pheasants and around seventy partridges. However, from time to time the reality differs for reasons which I will explain later. While we are definitely not professional game farmers, many of the techniques used at each stage of the process are the same as those used for rearing birds on a smaller or larger scale. The book takes you through each stage, from choosing and buying-in your own breeding stock, right through to releasing the end product for your own shoot.

There are several starting points in rearing game birds that are open to you. For example, you may not want to go through a breeding programme. This day and one where opinion should be sought from the professionals themselves. But perhaps the alternative title: 'game manager' is worthy of some discussion.

In the beginning we relied heavily on established techniques where we could find them, many of which rather confusingly conflicted. We learnt from these, and responded by progressively modifying our own approach to several of them. For example, we found that one of the highly sensitive areas in the whole cycle is the incubation process. There are countless opportunities for disaster here, yet it's really important to get that bit right. Accordingly, we have at times changed our tactics and find today that there are some pieces of published advice we just don't agree with. It's not a case of being bloody minded, far from it, especially since we were consciously incompetent at the beginning, and badly needed help. But our views have been born of honest questioning based on our developing knowledge. So, whilst we can never guarantee that our processes are fail-safe, they are mostly repeated and tested to confirm their workability. And we are always interested in new advice. In situations where we have departed from otherwise accepted techniques, I have outlined current published practice, explained how we do things differently, and why. That way you have a choice.

One of the principal lessons the whole rearing cycle teaches us is that it's not an exact science. That will certainly be one of the reasons for so many apparent disparities in the advice offered. Therefore, it's important to avoid jumping to lasting conclusions based upon the evidence taken from just one season and, incidentally, just one 'expert'. There are so many variables which can potentially produce different results. These range from the sperm count of the male, the health of the animals, extreme weather, trauma from predator presence, accuracy of measurements, etc., so it would be a brave person to put forward a set of definitive recommendations even based on a number of seasons of rearing. We are, after all, dealing with Mother Nature, not a scientist, and she certainly throws up several fickle inconsistencies at the most inconvenient times. So, at every stage of the process, the need for flexibility is essential.

Start Right

When we began rearing game birds our overriding objective was to ensure that, whatever the cost and effort, the health and welfare of the animals must come first. This is not meant as an opportunity to climb into the pulpit, it's just the way it should be. Our decision was, if we can't afford or manage to do the right thing, then don't attempt it. We've followed these principles ever since, and the mistakes that we have undoubtedly made have always been caused through incorrect information or inexperience, but never negligence.

A good sound starting point is the GFA (Game Farmers' Association) Code of Practice (2006 Edition), which outlines its Five Golden Rules. The complete document is well worth reading and has been a useful reminder to us, especially during periods of minor crisis like chasing around the yard trying to net yet another brooder hut escapee. After it has managed to squeeze through an impossibly small gap, we look on despairingly as it flutters around in wobbly circles trying to land. At times like this, of course, we'd prefer to leave it to that hovering hawk or – even if we could catch it – wring its neck and stick it on the barbecue – but sadly, that's not on!

The fundamental messages of this document offer sound practical advice, and some salutary reminders. Here are the Five Golden Rules, reproduced from the GFA Code of Practice 2006 Edition, with that body's kind permission.

1. Those responsible for captive game birds must be caring, considerate, conscientious, knowledgeable and skilled. They must be well prepared and take expert advice from veterinarians and qualified game consultants whenever necessary.

2. Captive game birds must have ready access to fresh water and an appropriate diet to maintain growth, health and vigour.

3. They must have an environment appropriate to their species, age and the purpose for which they are being kept, including correct heating, lighting, shelter and areas for comfortable resting.

4. Every precaution must be taken to avoid pain, injury, or disease. If they occur, diagnosis, remedial action and the correct use of medication must be rapid.

5. Captive game birds must be provided with appropriate space, facilities and company of their own kind to ensure the avoidance of stress.

Taking a responsible attitude towards raising your own game doesn't stop you from making economies where possible. Neither does it mean cutting corners; it's just taking a common-sense approach, and using your own facilities and equipment intelligently. For example we converted some old storehouses and dog kennels in a gravelled yard into brooder sheds. And instead of these opening out to a bespoke grassed exercise area, more conventionally used for rearing chicks, we put down concrete bases. More of that later, but this low-cost, easy to clean option has proved to be highly successful for us.

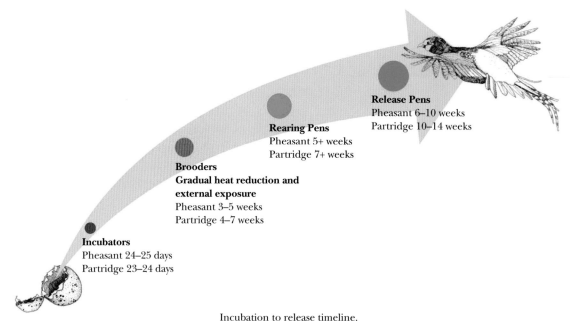

Incubation to release timeline.

Finally, just to provide a simple overview of what's involved, the adjacent chart shows the rearing process from scratch, to hatch and release. These timings are necessarily approximate and are dependent on many factors, several of which I have already mentioned. After that it's really up to you; the choices are yours. Just remember that the whole idea is to facilitate and enjoy your own shoot, a product of your endeavours.

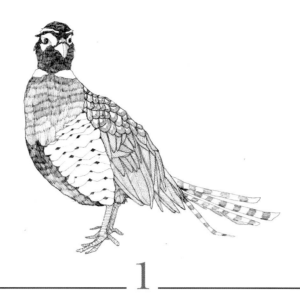

1

BACKGROUND AND BUYING-IN THE STOCK

'BRING IN THE BREEDERS'

There are several variables involved in deciding the best strain of game bird to rear. For example, do you want to admire it, shoot it, or eat it? Probably a little of each. But where you live, and the natural habitat available to the birds, may well influence your buying decision. A little background information is always helpful in assisting this process.

This chapter provides a brief overview of the birds' antecedents, and some examples of each species. The focus then shifts to the buying process and considers those issues associated with making your purchase.

Putting Pheasants and Partridges into Context

'Game birds' are defined under the Wildlife and Countryside Act 1981 as: 'any pheasant, partridge, grouse (moor game), black (heath) game or ptarmigan'. The emphasis in this book is on the pheasant and partridge.

In developing an understanding of these species we should take into account the important geological principle that 'the past is the key to the present'. It helps equip us with a clearer insight into their natural levels of resilience, adaptabilities, characteristics and instinctive behaviours.

Pheasants, partridges and their relatives are of the family *Phasianidae*. Members range in size from the domestic turkey to the diminutive quail. Interestingly, the red junglefowl, *Gallus gallus*, is also part of this family. It was domesticated around 5,000 years ago and is the ancestor to our domestic chicken. Together these species form part of the Order Galliformes, described as heavy-bodied ground-feeding domestic or game birds. There are well over 155 species classified in this group.

Pheasant History

The pheasant is a forest bird characterised by its long tail and, in the cocks, brilliantly coloured plumage. According to McGowan (1994), there are an impressive forty-nine species of pheasant and peafowl all of which, apart from the Congo peafowl, originate from Asia. Within this there are sixteen groups of pheasants, with the common pheasant and green pheasant falling under the same genus: *true pheasants*. Thereafter numerous sub-species exist (possessing natural, recognisable differentiation across populations), all of which are originally native to Asia.

To make any accurate attempt at tracing the early emergence of the species is difficult. It's true, there do seem to have been many different pheasant and partridge-like species present in Europe over 65 million years ago, although precise identification is unavailable. Those that date earlier, though, are thought to belong to groups which are now extinct.

More recently, it is probable that a fossil discovered in Namibia and dated at around 14 million years old could be a primitive francolin partridge ancestor. Confidence further increases with the discovery in China of early fossils of the common pheasant, which are dated somewhere between 1.64 and 0.73 million years old. So it is clear that these birds have been around for quite some time.

The native range of the common pheasant is vast: they existed at an early stage over much of Central Asia, Japan and China and their spread has been further extended by humans, almost worldwide. This has enabled them to successfully inhabit extremely variable climates and topographies. Many strains have emerged; a result of a combination of intermingled 'artificial' varieties and selective breeding. Some strains may have gravitated naturally to a particular terrain, but on the whole it's safe to suggest that their evolution in diverse environments has contributed to the pheasants' ability to flourish in a variety of conditions today. They have also been shown to adapt extremely well to aviculture which, it is thought, has been a further key to their survival over the centuries.

Whilst its native range is specific, the derivation of the pheasant's name is based on legend. The Latin name for the common or ringneck pheasant is *Phasianus colchicus* or 'the pheasant of Colchis'. This was an ancient country in the Caucasus, situated between the Caspian and Black Seas. Etymologically, the 'pheasant' has its roots in the Greek name 'bird from the Phasis', or the 'Phasian bird'. The Phasis was an ancient river known today as Rioni in what is now Georgia.

Legend has it that the pheasant, in this earlier habitat, was known to the ancient Greeks. Aristophanes and Aristotle, for example, make mention of it. It was then allegedly brought to Europe via Jason, who shipped it on the *Argo* after his search for the Golden Fleece in Colchis.

However, if you're not enchanted by Jason, you will probably accept some Roman history. It is more readily agreed that the Romans were largely responsible for bringing the pheasant to Britain, where it was used as a primary food source rather than for sporting purposes. Descriptions of rearing methods and food recipes have been found to support this view. Conversely, there's very little supporting evidence to suggest that the bird had actually naturalised in Britain at this time, and it's recognised that further study is needed to determine the physical evidence of the arrival of pheasants in Britain.

Further historical records of the species appear during the Norman Conquest, but again these are unreliable in terms of confirming actual naturalisation. Some time later, however, evidence emerges suggesting that the pheasant was used for the table during the reign of Edward I (1272–1307), where its value was about 4d.

We start to make more accurate progress thereafter. By the late fifteenth century there is finally proof that the pheasant was successfully established and breeding in Britain in the wild. Evidence of this is seen in royal records, wherein the birds were granted a degree of protection by the Royal Courts. Hunting and eating pheasant and partridge (together with most other creatures, as it happens) were popular with Henry VII's court. And Charles II is reported to have allowed only the richest 5 per cent of landowners to hunt anything at all, and imposed harsh penalties on lawbreakers. Some keen protective hunting instincts were clearly present even then! From that period on, managing to dodge the dining table from time to time, the pheasant gradually established itself, living freely in the wild throughout Britain.

So when you look at one of these game birds, consider its antecedents. Its remarkable ability to adapt to most habitats, including aviculture, is therefore not surprising given its even more remarkable history. For good reason, the pheasant has very quickly become one of Britain's iconic game birds.

The 'Modern' Pheasant

Over the centuries the British birds have interbred to become what they are today, a complete conglomerate of several sub-species. They exhibit variations in plumage colour, size and subtle behavioural differences. Whilst adaptable, individually these birds often suit different terrains and environments, and display different characteristics in flight trajectory.

With such a wide choice available, deciding on the right strain for you may seem hard. Nick Pardoe of Bonson Wood Game Farm says: 'Every strain of game bird will have its supporters. Generally speaking, on a lowland shoot, with little in the way of topography, a smaller wilder pheasant such as an American Kansas, Michigan blue or Bazanty will provide you with a more challenging target, whereas on a shoot with deep valleys a French ordinary or Chinese ring neck is perfectly adequate.'

Another variety to consider is the much larger Black-neck. Known to hold well, it's also good for the table, but critics believe it is a lazy flyer which reduces the sporting potential. Opinion is also divided about melanistics: faultfinders admit they are decorative, but believe they have a big tendency to stray.

I discussed this mixed feedback with Charles Nodder, spokesman for the National Gamekeepers' Organisation (NGO), who believes: 'Quality is generally more important than strain. Good flyers tend also to be wanderers. Shoots with big valleys can get away with a more docile bird which will hang around yet still be high enough on a shoot day. If you choose good flyers, expect them to be good walkers too.'

This commentary shows that there are many different views held on the same subject, often based on personal favourites. So avoid making hasty decisions based on the advice of just one person. We have a great choice in Britain, so you should be able to buy in the strain that best meets your needs. Talk to your neighbouring shoot; they will know what works in your area. And when briefing the game farmer, be clear about what you want so you can get suitable advice. Try not to become obsessed about the holding potential of each strain; this tendency may end up having more to do with your rearing management, game cover and keepering capabilities rather than the breed itself.

Less Common Varieties of Pheasant

To continue the introduction to the species, outside the usual group of game pheasants, there are three other cousins that do well in Britain, but for decorative purposes only. Zoe Hunter from Allandoo Pheasantry, breeders of ornamental and rare species of pheasant in Scotland, gives an excellent description of each.

THE GOLDEN PHEASANT

These are gorgeous birds endemic to the mountains of Central China [and brought to Europe around the fifteenth century]. The golden belongs to a group called 'ruffed pheasants' because of their 'cape'. In the golden pheasant this is black and yellow. The cock, once mature, with his wonderful array of colours, is one of the world's brightest birds.

They eat leaves, shoots, flowers, insects and spiders in the wild. In captivity they will do well on a proprietary game feed. They also enjoy many other treats, the favourite usually being peanuts.

Golden pheasants are often kept in trios (one cock bird and two hens) although sometimes a pair or even four or five hens can be kept. We do have a trio of golden pheasants which have consistently given us fertile eggs (about fifty to sixty a year) and have lived together harmoniously for a few years now.

Golden pheasant. *(Photograph courtesy Zoe Hunter)*

THE LADY AMHERST

It was first brought to Britain by Lady Amherst, but according to George Horne, the breed was properly introduced in 1869 by Mr Stone, of Scyborwen near Monmouth, who successfully managed to breed from his stock.

This pheasant is the second of the ruffed pheasants, endemic to China as well as Burma and Tibet. The cock Lady Amherst's pheasant has a beautiful long tail and gorgeous contrasting colours which include dark metallic green and blue with red feathers on the head above a ruff of black and white. The belly and lower breast is white, as is the tail, which is heavily barred with black. The tail also has shorter protruding bright orange feathers (upper tail coverts). The skin on the face will be a blue/green colour sometimes showing slightly yellow. Lady Amherst's pheasants are extremely regal looking birds, becoming fully coloured during their second year. The hens are sometimes confused with golden pheasant hens but they are bigger, with blue/grey legs and bluish orbital skin. Hybrids between the Lady Amherst and golden can be produced and are fertile.

A hen can lay as many as forty eggs in one season. Two or three hens are often kept with one cock, which can put less stress on each hen as the males can become quite aggressive towards them in the breeding season.

Lady Amherst's pheasants are very hardy and are easy birds to look after. They can have other birds in the same aviary, although no other pheasant species should be housed with Amhersts of breeding age.

Lady Amherst's pheasant. *(Photograph courtesy Zoe Hunter)*

We have owned breeding groups of each variety and, until they were efficiently despatched by a fox one dreadful night, they were a delight. In our first year we ran them all together with the ring necks and they were absolutely fine. The 'golden' is a lighter boned more restless bird. The Lady Amherst's are bigger and behave much more like typical game pheasants. Both are fabulous and, until their untimely deaths, made a rich and colourful addition to our adult group.

REEVES' PHEASANT

Originating from Central China, Reeves' pheasant prefers a habitat of open woodland. The colours of the cock may not be quite as bright or iridescent as some other species and he lacks a crest and face wattles, but his extra-long tail and ornate markings certainly make him a stunning bird. He has a white head adorned with a black choker and a second prominent black band surrounding the eyes like a Venetian mask. His body is mainly a shade of deep, dark yellow with a black border around each feather. He also has some white on his wings and a rich dark chestnut on the breast. The tail is mainly white with black barring.

Reeves' pheasant. *(Photograph courtesy Zoe Hunter)*

The Reeves' hen is a pretty bird, much paler than the cock but heavily mottled. She has a creamy yellow head with a brown cap and markings round the eyes to the back of the head.

The Reeves' pheasant does have a reputation for being aggressive, as documented by George Horne in his excellent book, first published in 1887 (from which I quote with the kind permission of Lightning UK). Writing for the amateur keeper, he says:

> It would be impossible to speak too highly of these noble birds, and those who have room will find but little trouble in rearing them; and to the game preserver I would say, what could be grander? They breed as easily as our common ones, and their great size and beauty render them most attractive. The more I see of the Reeves', the better I like them. I am told they are true 'rocketeers', and excellent sporting birds. Keep three or four hens to a cock. The cocks are very pugnacious and, when fighting, spring high in the air; they are also very bold in the aviaries or pens. I have had several which never allowed me to enter unless armed with a good stick, and I have had many battles.

And on his 'tour around his aviaries' he reports:

> We now come to a pen of Reeves'. Here I dare not enter without a switch, for Beaconsfield, as we call him, because his motto is *semper paratus*, at once prepares to drive me out. We have had many fights: once I stunned him by accident, and once he slipped out after me to attack me, darted at me, and then flew away. Here was a fix.

Whilst not all of Mr. Horne's animal management techniques are likely to meet today's various inexhaustive Codes of Practice on the subject, his accounts and advice do make entertaining and often highly educational reading.

Happily, not everyone declares this magnificent bird an official thug. Zoe Hunter gives a different account altogether. On the question of temperament she states:

> We only started breeding Reeves' pheasants in 2009 as we were rather put off by the aggressive reputation this bird has earned. We seem to have been lucky, however, as we have had no problems with our adult Reeves' at all. … Our birds have certainly shown us they are not *all* villains and should definitely be given the chance to prove themselves worthy.

The Common Pheasant *Phasianus colchicus*

In terms of biometrics the cock pheasant measures typically between 74 and 89cm (30–35in) long, with a tail length of 42–59cm (16–23in) and weighs on average 1.4kg (3lb), whilst the hen pheasant is smaller, being around 53–63cm (21–25in) with a tail length of 28–31cm (11–12in) and a weight of around 957g (a little over 2lb). They have a wingspan of 70–90cm (28–35in). Basic colouring for the cock bird is vibrant. The head is usually blue green with red face wattles and a distinctive white neck ring. The overall plumage is chestnut brown, with golden-brown, black and pale blue markings on the body. The tail is chestnut brown with black bars. The lower back varies from chestnut bronze to pale blue and purple. Some have silver flanks and others are midnight blue/black, which identify different strains.

Two common pheasant cocks, showing colour variation.

A common pheasant hen.

Hen birds are more uniform in colour and exhibit an excellent camouflaged plumage of light and dark browns.

The cock sports the celebrated magnificent long tail whilst the tail of the hen is rather less flamboyant, nevertheless ending in a regal point.

Characteristics

If you are an experienced shot you will know that these birds are most comfortable on the ground. They can, of course, fly and when spooked will launch themselves airborne with an abrupt, noisy take-off exploding into the sky in a 'flush'. They are apparently capable of reaching speeds of up to 96kph (60mph) over short distances, but we have never witnessed this. The distinctly reduced speed of 48–64kph (30–40mph) is probably closer to the mark. More usually they will run from trouble, and can often be seen sprinting to safety with great alacrity, hence the benefit of using a team of beaters in a driven shoot, who literally flush them out of cover and into the air. Generally their preferred habitat is fields and farmlands with brushy cover, although being adaptable birds they also inhabit woodland undergrowth and some wetlands.

In terms of their ecology pheasants can be found alone or in small flocks.

They tend to form single-sex flocks during the winter, although this depends a little on the availability of a suitable habitat. The hens prefer woodland with shrubby undergrowth, and the cocks gravitate towards slightly more exposed areas, like hedges and woodland edges. Hens disperse from their winter flocks during late March and early April. Cocks also move out into more open areas, taking up territories which encompass the edges of woodland.

Cock pheasants are polygamous and establish harems of up to a dozen hen pheasants. This can be a tricky job since hens have a habit of continually joining and leaving a group. Each spring a cock defines and defends his territory and his harem from unwanted rivals. Such encounters can lead to apparently vicious battles, although they usually amount to 'handbags at dawn', with a great deal of squawking and posturing followed by the occasional lunge, but rarely resulting in a bloodbath. Once his patch is established, a cock bird will focus his efforts on attracting and then retaining his retinue of hens.

Both the cock and hen reach sexual maturity in their first year. The hen normally lays her eggs between April and June. Typically egg size is around 46×35mm (approx. $1\frac{3}{4} \times 1$ in). The eggs are smooth, either glossy or non-glossy, and are usually olive/brown. There are departures from this norm, as I will describe in Chapter 2, which we have found to be acceptable differences from a hatchability point of view.

The hen's preference is to nest in fields or hedge/woodland border habitat where she can lay at least a dozen eggs. If disturbed, she will often abandon the clutch and may make up to four more nesting attempts. The incubation period is 24–25 days, although our experience using artificial incubation processes has, at times, produced different hatch times.

The hen incubates her eggs alone. In ornithological terms chicks fall into one of two scientific categories, each of which describes the degree of development in young birds at hatching. Pheasant chicks are precocial, which means, as with many other species such as chickens, ducks and turkeys, they mature very early. They differ from altricial young, e.g. those of owls, hawks and woodpeckers, which form the other broad grouping. This latter group of chicks are incapable of moving around on their own immediately after hatching and need nourishing.

Born with their eyes open, pheasant chicks are covered in downy feathers. Sometimes feathers of the juvenile plumage may already be present, such as the first seven primaries. The other primaries appear during the following three to four weeks. The first moult begins at about twenty-three days old. It takes 150 days for the primaries and secondaries of the first winter plumage to complete their growth, and 170 days for the completion of all the coverts. The tail feathers grow more rapidly.

'Officially' pheasant chicks can fly within two weeks of hatching, but 'flight' at this stage generally amounts to an ungainly flutter and, in our experience, careering into brooder nursery walls. In the wild they will remain with their mother for six or seven weeks, typically until early autumn, during which time they will be taught basic survival skills.

Young pheasants moult their juvenile feathers at seven weeks, and begin to develop adult plumage and coloration. Family groups begin to break up shortly after the brooding period and are completely dispersed by September. Cock pheasants do not exhibit a full display of colours and the characteristic white neck ring until they are about eighteen weeks old.

The Partridge Family

Partridges are present throughout much of the world and, as with their pheasant cousins, their native range is highly diverse, although some partridge species do favour specific conditions. For example the grey-breasted francolin *Francolinus rufopictus* has a native range of the tropical open country of East Africa, the Tibetan snowcock *Tetraogallus tibetanus* comes from the Pamir and high alpine zones of the Himalayas, and the native range of the red-leg partridge *Alectoris rufa* is south-western Europe.

Through the ages, as with the pheasant, the partridge became a popular choice for the table and was later recognised as a challenging sporting bird. There are several sub-species, each of which possesses slightly different characteristics. But, on the whole, they can be safely described as a rotund bird that prefers to run rather than fly. If they do take to the wing they generally travel short distances only, and flap rapidly, creating a whirring noise which is common to the species. They may also glide as part of their flight pattern.

The Grey Partridge

Probably our most well-known partridge is the Grey Partridge. To describe the species I have used with their kind permission words of the Game & Wildlife Conservation Trust (GWCT), a world authority on grey partridge research:

> The grey partridge was originally a bird of temperate steppe grasslands. It has adapted readily to open arable landscapes and, accordingly, vastly expanded its range as agricultural development spread westwards across Europe over the last

eight millennia. After the last Ice Age, the grey partridge arrived naturally in Britain (and was first formally recorded in Britain in the eleventh century).

The combination of land enclosure, increased cultivation and intensive predator control in the eighteenth and especially the nineteenth century boosted its numbers considerably. It emerged to become the most popular sporting quarry of the last century. Bag records show that the largest numbers were shot between 1870 and 1930, during which period around two million grey partridges were killed annually.

The same bag records indicate that, after the Second World War, the numbers of grey partridges dropped by 80 per cent in 40 years. Our research has established three main causes for the decline:

1. Chick survival rates fell from an average of 45% to under 30% between 1952 and 1962. In the first weeks of life, grey partridge chicks feed almost exclusively on insects to obtain the proteins needed for rapid growth. The introduction of herbicides in the early 1950s eliminated many crop weeds that were insect food plants and, by the 1980s, the number of chick food insects in cereals had fallen by at least 75%. Although the drop in chick survival rates was partially compensated by lower over-winter losses, it reduced autumn stocks sufficiently to upset the economics of game management.

2. Many gamekeepers either lost their jobs or turned towards pheasant rearing, resulting in less predator control and an increase in predation during the nesting season, leading to more hen and nest losses.

3. In some areas the situation was exacerbated by the removal of grassy nesting cover as fields were enlarged by removing hedgerows and field boundaries.

Today in Britain the greatest densities of grey partridge are found in the lowland arable areas of the south and eastern side of the British Isles. However, they also extend into the north of England, and east of Scotland as far as Aberdeenshire. Scant numbers can be found in other parts such as the north Pennines and Midlands, but they are very scarce in Wales and Northern Ireland.

In terms of biometrics, the grey partridge is 28–32cm (11–13in) long with a wingspan of 46cm (18in), and weighs around 0.33kg (¾lb). It is a plump bird with a relatively small head. It has a reddish face and tail, grey breast, barred sides, and a dark reddy chestnut colour 'U' shape on the belly. Both sexes look alike, although the hen is slightly smaller and duller, and the belly patch is reduced.

Grey partridge. (*Photograph courtesy Laurie Campbell*)

Many gamekeepers will use both grey and red-leg partridges (see next section) on the same shoot. Their logic for this varies but, on the whole, it is agreed that the grey partridge shows excellently but is a canny, evasive bird that can be highly challenging for the Gun. Some keepers also believe they have a stronger wandering instinct than their cousins. The combination of the two often results in a better shooting return from red-legs. As with all partridge strains, they form and stay within sociable groups called 'coveys'.

Sedentary in nature, grey partridges breed and winter on farmland, grassland and arable fields. The hen usually makes one nest but may have a second attempt if disturbed. She lays an average of fifteen eggs (but can produce more), on the ground in a grassy cup in amongst tall plants or hedges. The eggs, measuring around 36 × 27mm (approx. 1 × 1in), are smooth and glossy, and uniform olive brown. The hen alone incubates her eggs. The precocial nestlings are able to feed themselves, but both parents care for them. The chicks' wing feathers appear after about five days and they can flutter at ten days, but they are not fully fledged until twenty-eight days. Juveniles are brown, have yellow legs and lack the orange throat and dark belly patch.

Other Family Members

To the amateur, several of the partridge family members resemble one another closely but are distinctly different from the 'grey'. The Rock partridge *Alectoris graeca*, Barbary partridge *Alectoris barbara* (the national bird of Gibraltar), Chukar partridge *Alectoris chukar* and the red-leg partridge *Alectoris rufa* broadly speaking all feature a light brown back, rufous-streaked flanks, grey breast, mainly buff belly, dark collar and red legs. There certainly are differences, especially in shading and shape of gorget, but as a group they are similar and much gaudier birds than the grey. Interestingly outside Britain, the Chukar is a very popular game bird, especially in North America, where it thrives in dry, arid conditions.

This simple overview provides a glimpse of some of the less familiar members of the species. It is unlikely that you will see many of them, but I include them to give our more popular sporting game birds some context in terms of their relations.

The Red-leg (French) Partridge

The red-leg partridge *Alectoris rufa,* also colloquially known as the French partridge, is not an indigenous species to Britain. It measures between 32 and 35cm (12½–14in) long, and has a wingspan of around 47–50cm (18½ –20in). It may have been introduced into Britain by the Romans from continental Europe, and the first known records point to introduction from Spain. It is certainly the case that these birds were around during the reign of King Charles, who using broody-reared birds from France for shooting purposes. Charles was anxious to establish the birds to compensate for the falling numbers of the grey partridge, which had resulted from over-hunting. The red-leg population expanded, especially with the rise in popularity of hunting and shoots.

The red-leg is a larger, more colourful version of its grey cousin. Despite this it is still well camouflaged. Red-legs blend especially well into their preferred habitat of heaths and downs, and thrive best on dry, sandy soils. The red-leg has a large white chin and throat patch, bordered with black. It has a greyish body with bold black flank stripes, and a chestnut-sided tail. Whilst similar to its other cousins, it is distinguishable from Rock partridges and Chukars in particular, by the broken spotted black collar on the neck and throat.

Feeding habits are largely common to all of the species. They eat mainly leaves, roots and seeds of grasses, cereals and weeds, and insects – especially when feeding chicks where a starter diet of insects and grubs is vital. They usually nest among bushes in scrub, arable farmland, or hedgerows. As with

Red-leg partridge.

that of the grey partridge, the nest is a shallow hollow on the ground, lined with plant material.

Red-legs breed most successfully in hot summer temperatures, but may begin much earlier from late April, warmth and dry conditions being an important combination. They typically lay ten to sixteen eggs, but can lay more. The eggs are smooth and non-glossy, yellowish-white with reddish-buff or greyish markings, measuring approximately 41 × 31mm (1⅝ × 1¼in). The incubation period is twenty-three or twenty-four days.

They can sometimes exhibit a strange nesting behaviour, whereby the hen builds two nests, laying a clutch in each. The cock then incubates one and the hen the other. We have experienced this several times and at first thought it was an accident of Nature. It is definitely an idiosyncrasy that demonstrates a highly resourceful way of dealing with a larger brood simultaneously.

As with pheasants, the nestlings are precocial and are led away from the nest shortly after hatching. The wing feathers grow quickly and they can flutter quite early – so if you decide to rear them for the first time, beware – they're like bullets, especially when they ricochet off netting! The brood remains together until the following breeding season.

In the British Isles they are most commonly seen in England and, as with the 'greys', are most prevalent in the east, with some birds also in the Welsh borders and in eastern Scotland. You can normally spot them in coveys in open fields.

As a sporting bird, red-legs are a fast and strong-flying bird. They hold nicely in cover and produce well when flushed. They use their deep chests to propel themselves uphill and flush downhill, often giving Guns numerous chances. They also make good sporting targets in flight, owing to their habit of flying in small groups of two or three birds. As with the greys, they are great sprinters and scurry across the ground at high speed to avoid predators. On the whole they are considered by many seasoned gamekeepers to be 'a good looker, and great all-rounder'.

Getting Started with Rearing

Once you decide which game bird species you intend to rear the next decision is to work out where you want to start. I mentioned this earlier but it's genuinely worth stating again. Don't over-face yourself and become too expansive. Begin at the stage where you know you have the time, site, equipment and funds available to manage it properly, otherwise you will end up resenting the whole project. That's not fair to you or the birds. Happily there are options, none of which could ever be considered 'a breeze', but some are definitely less resource-consuming than others.

There are several starting points, each with their own advantages and disadvantages. Once you've established the foundations of your stock, you essentially repeat your options with the following possibilities. You can retain a 'closed flock', which is a group of parents that are penned in and separated from any others; this allows you control over their health, welfare and breeding. Alternatively, you can catch-up birds left at the end of the season and keep the healthy, calm stock to use for next year's rearing cycle. If you do decide to catch-up, or have a combination of both systems, it's always better not to mix the birds immediately, since this risks introducing disease to the closed group. If possible, segregate the wild stock and take them through a comprehensive parasite control programme. Assuming there are no issues with the species, you can then mix them in later. The other option is to rely on birds reproducing in sufficient numbers in the wild to sustain a future shoot – however, supplementary stocks are normally necessary.

Sourcing Stock – General Guidelines

In describing the main 'starter' options I decided to seek advice from some experts who have years of experience in buying, selling and keepering game birds. I am extremely grateful in particular to Bettws Hall Farms, Bonson Wood Game Farm, the Game Farmers' Association (GFA) and MacFarlane Pheasants for their advice.

The first job is working out how to select the best-quality game farm. It's always good to support your local community and choose a farm that's close. Clearly that's also preferable for the birds, but it's not always possible. If your neighbouring shoot favours a particular supplier, ask their opinion – you can't beat references. If the neighbour is very close and the shoots border each other, it may also be sensible to use the same game farm. The birds will eventually mix and, if they are from different sources, diseases to which some birds have become immune could be passed on to birds whose immunity has not yet been challenged.

If you draw a blank there you can always take advantage of the wide variety of helpful organisations that will be able to point you in the right direction. Farms which are members of the GFA, for example, are obliged to follow the relevant government welfare codes, and have to provide a reference from their vet before they can join.

If you decide to crack on and do it yourself, try to avoid judging the quality of a game farm solely on the basis of a telephone conversation or flashy leaflet. The best thing you can do is to visit and check out their protocols, bio-security practices and the birds themselves. If you have a knowledgeable colleague, take them too. Consider also the length of time the business has been running. If it has been trading for many years, it can't be doing too much wrong! If satisfied at the end of the visit, take up references just to be sure. Your investment at this stage is too great to take poorly researched decisions.

If you do visit, when you are shown the stock, some basic things you should be looking for will be evident from the general appearance of the birds. I don't suggest you should whip out your stethoscope at this point; just observe keenly, and use common sense. Are the birds thin and withdrawn, or lethargic? Do the wings droop and are their heads held lower than normal? Unhealthy birds tend not to move much. They retract their necks a little, are hunched up and often have ruffled feathers. Does the whole flock look like this, or is it just the odd bird? If you spot any of these signs, ask about any current health problems and whether treatment programmes are in place. If the answers are 'No', and 'None' respectively; leave quickly and disinfect your boots when you get home!

If possible, have a bird caught up for you to scrutinise a little closer. Look for parasites crawling on the bird. Do not buy anything crawling with lice, these are treatable but are an indication that the bird has other health problems, which is why the parasites have had the chance to become abundant. Scaly legs or scummy eyes are also signs of disease.

Observe the birds walking about. You are looking for birds that have good strong, smooth legs, straight toes and that walk well. Some birds seem to high-step when they walk, which is not good. For red-leg partridges, look for a nice strong red colouring in the leg. Wings should be held close and high up. Full feathering with a healthy sheen, a sharp eye, together with a 'normal' posture and overall balanced behaviour, are all winning signs in game birds. And finally, take a glance at the feed and water receptacles, as well as the bedding; do they appear to be clean and fresh? If neglected, this is not the game farm for you.

So, although the ideal is to choose a local supplier, don't worry if you can't. Bear in mind that game birds are usually transported overnight to avoid undue stress, so your theoretical safe geographical limits are potentially greater than nationwide.

Four Buying Options

These are:

1. Adult breeding stock
2. Eggs
3. Day-old chicks
4. Poults

Adult breeding stock

These birds can be bought 'good to go', or intended to be; depending on the time of year you buy them in. Generally speaking, birds can be obtained at poult age (see later) so you can over-winter them, and use them as parent birds the next year. Another buying window is after the shooting season in February, when the keepers are catching them up. Just be aware that most of these birds will have been shot at, so may be a little jumpy. It may also be possible to buy mature stock. If these are available you should buy them in the autumn so they have a decent period to settle in before the start of the next season.

Eggs

These are only an option if you already have the incubation kit up and running. So, if this is your choice, there are several considerations and questions to ask. Find out whether the eggs come from a closed flock or not. Clearly the former is better because their health and welfare *should* have been closely controlled. Ask for the history of any disease outbreaks that could be transmitted via the egg. Linked to this is the question of whether the game farm is able to trace the eggs back to the batch of hens which have laid them. This is important because you will want to know whether the birds have come from proven flying stock and, importantly, what the strain is.

Other hygiene-related questions include finding out about the system used for cleaning and disinfecting the eggs. Eggs from GFA member suppliers will all have been washed on the date of lay using a bespoke wash powder which cleans and protects the egg. Ideally, they should also be fumigated prior to set and despatch.

You can also ask whether the laying system is on or off the ground. A modern raised laying system is obviously cleaner, and helps reduce bacterial build-up that can sometimes occur where eggs are laid on the ground. If not, ask to see the nesting pen.

When discussing the number of eggs to buy, ask for hatch success rate probabilities. Guarantees are hard to give, so ask for previous history based on the same parentage. If results show a bad hatch then you may have a case to be supplied with more eggs or chicks, or a partial refund.

You will also need to know when the eggs were laid and the length of time they have been stored before despatch. Eggs can be stored for up to fourteen days, but the longer the storage, the less fertile they are likely to be. Make sure you have the eggs boxed with the date they were laid written on the box. That avoids any confusion. Once received, set the eggs as soon as possible thereafter, having allowed them a reasonable 'settlement period'.

Day-old chicks

These might be a good choice if you're already set up to receive youngsters, but beware of false economies. Just because they're cheaper than the adults and poults doesn't mean that, because of equipment and time requirements, in the long run you won't be spending more. Realistically, you should also factor in a loss rate of around 18 per cent.

If you decide on this option, ask about the hatch percentage. If there has been a bad hatch, then the chicks may well be substandard. Bad hatch rates often result from some external factors which were wrong during incubation.

Ask how the breeding flock was fed and managed. Find out what the strain is, and the hatch date. Pheasant chicks can be sexed at day-old, so you may want to state your number preference for each gender. Ask the supplier where to buy feed, and what medication, if any, to use. Never skimp on food quality; always buy the best you can. The supplier should also tell you if the chicks need to be medicated for any problems the hatchery has been experiencing throughout the season.

If chicks are boxed and despatched soon after they hatch they can travel for up to thirty-six hours without harm. This is because they live on the yolk of the egg, which is drawn through the navel into the stomach. However, the transport truck must be suitably heated (or air-conditioned), and ventilated. Nevertheless, time is important, so find out what the delivery details are. If you have any real concerns you should arrange to pick up the birds yourself, or choose a supplier that is closer to the shoot. Normally a few extra chicks are supplied as part of the contract. This is common practice: just find out what the percentage is so you know exactly what you're paying for.

Poults

Pheasant poults are normally sold at between six and nine weeks of age. Any younger and they will not be properly feathered, and therefore unable to withstand adverse weather; any older than nine weeks, and you may have difficulty holding them in the release pens. Partridge poults are usually sold between ten and fourteen weeks old.

When buying-in you can decide on the number of each sex you want, but they are usually sold as 50 per cent of each gender. The supplier should be clear as to the exact age of the poults and tell you if there have been any health issues with the birds. If you want the pheasant poults wing-clipped this should also be an option, but usually at an additional cost.

If the birds are going to go directly outside in the release pen, despite any assurances that they are hardened off, try to plan your delivery to coincide with a period of two to three days without rain. A tall order, I know – but it is a serious point.

Catching and transporting poults is extremely traumatic for them. The unavoidable stress this causes leaves them vulnerable to disease. It is therefore important to make the transition as easy, gradual and as familiar as possible. Find out exactly what food, medication and type of drinkers and feeders have been used. Several shoots use nipple drinkers, for example. They are seen by many as a huge help in reducing disease transfer. But if the birds have not come across them before, it may be a good idea to rent a few drinkers from the game farm and fit them alongside your own. The other alternative is to provide the game farm with the drinkers that you want to use so that they can be rigged up in the rearing pens. Exactly the same principles apply to the introduction of feeders and food. The birds' gut flora needs to adapt to any new foodstuffs, so introduce changes gradually.

Delivery

Being subjected to transportation is extremely stressful at any age of the game birds' life, so the more you can do to minimise the trauma the better it for them. Unfortunately, fatalities are not uncommon, but if you deal with a quality game supplier, then any losses should be considered an exception. Think about everything you can do to reduce the birds' stress. Access to the pens is an obvious point – can the delivery driver get close, or ideally drive straight into the release pen? This avoids the birds having to be hiked around or subjected to yet another journey, which also further extends human contact.

A good-quality game farm will issue a vet's certificate that states that the birds were fit at the time of departure. Ask for one. On arrival you will usually be asked to sign a document confirming that the birds are healthy and as per your order. Paperwork details should include dates, number of crates and number of eggs or birds per crate, time and date of catching, condition of the birds, what feed they have been on and a list of past medication. It should also tell you where the birds came from, their strain and age. Some documents also include the driver's name, weather on delivery and comments about the condition of the pen to which they are to be delivered. Don't be frightened to give feedback straight away, whether positive or negative.

When making your order you need to be content that the supplier is using the correct type of carrier, and understands how many birds can be loaded safely into each. Ask when the birds were loaded and what number 'drop' yours are. Often, birds are transported overnight. This is infinitely preferable as it reduces disruption to their feeding routine. Always ask for an early morning delivery if possible, but have some sympathy with the driver dealing with our great British road system and traffic. Make sure you have a contact number and get the supplier to call you if there is any alteration to delivery times.

If none of this works for you, it is usually possible to pick up the birds yourself. Be prepared though: a reputable game farm will disinfect your vehicles and crates and may ask you to take some bio-security precautions. This may be a bit annoying, but these precautions are positive signs that demonstrate the quality of supplier.

Our Initial Experience

In this section, I'll explain what happened to us in the early days. Obviously, everyone's starting point and circumstances are likely to differ in some respects, but our experiences may provide a useful insight into certain issues.

First purchases

Our first stock purchases were a combination of parent breeding birds and some poults to give us an early show in the woods. We kept some of the poults back and they became parents for the following year. This worked well and was a cheaper option. To buy-in eggs or day-old chicks was not an option for us at that stage because we did not have the equipment needed.

A word on numbers: always buy more stock than you think you need. You can usually expect a few extra as part of the deal, but that will only help at the beginning of the process. The sad facts of life are that, with a combination of disease, injury, predation or simply escapees, despite your best efforts losses are inevitable. We bought-in 20 per cent more stock than we thought was necessary. That worked well, and was needed.

Containing the breeders

Once we had ordered our small breeding population of pheasants and red-leg partridges we needed to find somewhere to over-winter them. As I explained at the beginning, our objective was to use the existing facilities where possible before spending money unnecessarily. We had inherited two possibilities. The first was a good-sized covered pen measuring approximately 50 × 30m (165 × 100ft), with an interior split into three sections: one large area with two integral, smaller pens, each having a gate and hen coops incorporated. The second option was a simple rectangular aviary that was probably used years ago as some kind of mini-release pen. This was partitioned down the middle, with two gates providing two separate enclosures. Whilst none of this was pretty, we knew we had the basics. The structures were solidly built and suitable for our incoming game birds.

The first job was to make the pens habitable. As with all aviary constructions, we needed to make sure they were robust and predator-proof. Starting from the ground level soil line we checked that the wire mesh was flush to the earth all the way around. Neither pen had been constructed with the wire netting properly dug into the turf, so we fixed planks 2m × 20cm (6ft 6in × 8in) around the baseline, and pegged them in using 40 × 7cm (16 × 2⅜in) wooden wedges. The pens were double-layered with a combination of a 2m (6ft 6in) high galvanised main chain-link mesh, with 50mm (2in) diameter holes, and smaller 30mm (1¼in) diameter wire netting wrapped around. This was a good start, but only a tiny level of close inspection revealed many rusted areas and several holes. Unsurprisingly, there were also countless rusty nails and other unpleasant protrusions, as well as broken perches. After nearly a week spent patching them up, we did start to wonder whether, in the long run, it might have been easier to build a couple of new pens. But with cost concerns, and reluctance to waste otherwise decent structures, we soldiered on.

One of the inner pens contained an ungainly grouping of ancient hen coops and the other a lean-to, under which was a spectacular assortment of ancient wooden boxes firmly nailed together by a huge lump of oak. Well, it was a start. One of the things we were extremely conscious of even at this very early stage was the importance of hygiene. Parasites and bugs are very capable of living for a number of years in the soil and woodwork so, as a precaution, we stripped out and burnt anything that looked too ropey and soaked the rest with avian-specific cleaning products. We then double-dug, rotovated and re-seeded 80 per cent of each pen. To the final 20 per cent we added a mix of grit, sand and stones. This is designed to help the birds with their digestion and provide a firm base to walk on in wet weather. Our final touches were the addition of natural shelters (the detail of which is covered in Chapter 5), perches, feeders and drinkers.

Arrival

Delivery day for our birds finally dawned. We had selected a game farm that appeared to have a good reputation, but we had not at that early stage gone into the research process in as much depth as I suggested earlier in this chapter.

The company promised fully air-conditioned transport, with birds contained in bespoke carriers. What arrived was a van with lots of holes in the sides and our game birds stuck in strange-looking boxes. Even more unsettlingly, they had shared the journey with several other creatures which were bunched up in odd-looking crates, many of which were definitely not pheasants or partridges, and one scant document. However, to our relief, they seemed to be unscathed; apart from one hen poult that was very obviously dead. We quickly transported them to the pens, separating the poults from the breeding stock. They inched their way out of the carriers gently and immediately began foraging.

With the benefit of hindsight I now realise how naïve our buying process was, and that we probably got away lightly in ending up with such a robust group of birds. As you can see, nowadays the sale of livestock is fairly well regulated, so errors of judgement like this should be avoidable. But don't do what we did; make sure you ask several more searching questions before placing your order.

Routine husbandry jobs

We quickly settled into the job of managing the stock. Just as with humans, routine is important to the birds. We check them a minimum of twice each day to make sure that food and water containers are full and clear of debris and droppings. They actually seem to enjoy these visits and gather round to inspect any titbits we bring along. The partridges and hen pheasants in particular love greens and will greedily peck at bunches of alfalfa or lettuce, bits of courgette, etc. It certainly seems the way to their hearts.

Much is written about the temperament of each species and in our experience the cock partridges and pheasants are relatively composed. But the hens can be a little different, especially the partridges. Even a sudden change in colour or style of clothing can really put the birds off, so don't be tempted to 'cut a dash' with a bright outfit in your pens; 'dowdy' is the acceptable order of the day. We haven't had much trouble with the wardrobe side of things, but we have with the arrival of strangers. A sudden influx of humans does seem to upset the avian equilibrium, inevitably having the greatest negative impact on the more nervy hens. This is definitely compounded when the new arrivals are noisy. Coupled with that are the occasions when we do our maintenance chores outside of the routine time periods. This is definitely observed with some suspicion, and a threatened attack of hysteria from the partridges who are now convinced we are up to no good.

So, through trial and error, we found the best way to handle the birds is to apply common sense and:

1. Try to keep routine visits to roughly the same time each day.
2. Move around quietly amongst the birds. Talk gently and avoid flamboyant body movement or speed.
3. The addition of titbits is always a welcome 'sweetener', but never just discard them on the ground – this simply encourages the pecking instinct and with that comes the potential ingestion of parasites, an occupational hazard with ground-feeding birds. Find a spot to thread titbits in the netting, or use a bespoke area that is easy to keep clean.
4. If you do have visitors, or indeed the vet coming to look at the birds, be firm in issuing house rules. Don't be at all embarrassed about this; these are the foundations of your future generations of game birds. Rules should definitely include no screaming children, no running around or extravagant body movements, and keep numbers to the bare minimum inside the cage at any one time. Visitors should also adhere to strict hygiene safety rules that include dipping the soles of their shoes in a disinfectant bath and using a hand cleanser. That usually puts a few off!

Visits should also be treated as an opportunity to closely observe your stock and surroundings. Admire them by all means but look out for signs of bullying, feather-pecking, injuries or off-colour birds, as well as assessing the pen construction itself. Check to see that the netting and other materials haven't suddenly sprung holes or come loose. This will be of special relevance if you decide to do a renovation job like we did. Check also for predator activity. It need not take long – just incorporate these basic observations into your maintenance visit which, in any case, you are stuck with.

Winter and spring

One of the things we've had difficulty with is the impact of inclement weather on the birds. We have heavy clay-based soil which, in the dry weather, becomes incredibly hard and cracks, and can play havoc with the pen foundations unless the fence base is dug in deep enough. The soil isn't too bad for the birds in the dry, but because of its natural constituents it does lack grit and stone. This is easy to fix, though, with a mix of grit and stone chippings scattered in designated spots.

If you have this soil type you will know that the worst aspect of it is the way it changes in wet weather. It transforms from a hard-packed dry, dusty base into a claggy, viscous quagmire. Because of its slow water absorption rate it becomes extremely difficult for the birds to keep clean. Once the grass has died off, it is not uncommon, during bouts of very wet weather, to see the birds with mud halfway up their legs. This inevitably leads to the risk of disease and parasite infestation as they peck to get rid of the stuff.

Having suffered some health scares from this in the early years, we now manage these periods of inclement weather much more successfully. Living in a fruit-farming area we beg a number of old pallets from neighbours who are very happy to donate if there's the promise of a brace of pheasant or partridge in return at the end of the season. We place them at intervals inside the perimeter line, and a couple in the middle. This helps prevent muddy tracks appearing as the birds patrol up and down the sides. As it happens, they don't seem to want to escape – it is just a peculiarity. In fact we have on occasions inadvertently released a 'breeder' only to find it has followed us straight back in again having experienced a sudden attack of agoraphobia. Incidentally, we drop seed in between the pallet slats to encourage ground-cover re-growth. We have also laid weed sheets with a thick layer of gravel on top on known really wet spots.

If we have bouts of extremely wet weather we will also put down deep litter, using straw, hemp or wood shavings. Many veterans may vehemently disagree with this practice, warning against it because of the potential build-up of mould

spores. But so long as you are prepared to remove the whole lot, including the inevitable muck beneath, at the end of the wet period, you should be fine. And frankly, it's better than the alternative.

We also put feeders and drinkers on masonry stone chippings or pallets. The type of feeders we use have an iron rod stuck into the ground with an integral metal hoop on top (rather like the frame of a netball goal), into which the plastic feeders are placed. These are basically a bucket with a lid and detachable metal mesh basket that slots into the base, which contains the food. The birds then peck at the basket, which releases the grain. We put either a plank or cheap paving stone underneath it – basically anything that avoids them feeding directly on the wet turf and which can be removed and cleaned easily when necessary. These very basic tactics are designed to keep the birds comfortable and disease to a minimum. And we are certain they have contributed to the maintenance of a healthy flock of game birds and the resultant production of many fertile eggs for several years.

2

EGG-LAYING AND INCUBATION

'THE PRODUCTION LINE'

This chapter deals with the egg-laying and incubation process. We have recorded our results each year over the past four years, which is short by scientific study standards, but our conclusions are detailed and based on empirical results. The focus is on using machines as the main tool for incubating eggs. We do also use 'the real thing', and surrogate mums, but only to a minimal extent which I outline in Chapter 5.

The Egg

When we started out we found it helpful to understand, at a basic level, something about the make-up of the egg. This is especially useful when you are involved in an assisted birth:

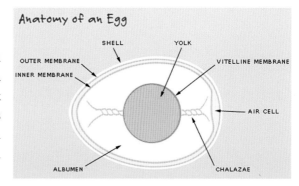

Shell. Bumpy and grainy in texture, an eggshell is covered with as many as 17,000 tiny pores. Eggshell is made almost entirely of calcium carbonate ($CaCO_3$) crystals. It is a semi-permeable membrane, which means that air and moisture can pass through its pores. The shell also has a thin outermost coating called the *cuticle* (*bloom* in chickens) that helps keep out bacteria and dust.

Inner and Outer Membranes. Lying between the eggshell and egg white, these two transparent protein membranes provide efficient defence against bacterial invasion. If you give these layers a tug, you'll find they're surprisingly strong. They're made partly of keratin, a protein that's also in human hair.

Air Cell. An air space forms when the contents of the egg cool and contract after the egg is laid. The air cell usually rests between the outer and inner membranes at the egg's larger end, and it accounts for the crater you often see at the end of a hard-cooked egg. The air cell grows larger as an egg ages.

Albumen. The egg white is known as the albumen, which comes from *albus,* the Latin word for white. Four alternating layers of thick and thin albumen contain approximately forty different proteins, the main components of the egg white in addition to water.

Chalazae. Opaque ropes of egg white, the chalazae hold the yolk in the centre of the egg. Like little anchors, they attach the yolk's casing to the membrane lining the eggshell. The more prominent they are, the fresher the egg.

Vitelline Membrane. The clear casing that encloses the yolk.

Yolk. The yolk contains less water and more protein than the white, some fat, and most of the vitamins and minerals of the egg. These include iron, vitamin A, vitamin D, phosphorus, calcium, thiamine, and riboflavin. The yolk is also a source of lecithin, an effective emulsifier. Yolk colour ranges from just a hint of yellow to a magnificent deep orange, according to the feed and breed of the bird.

Image and information, ©Exploratorium, used with their kind permission.

Laying

Natural versus Artificial Incubation

Artificial incubation is discussed in detail later in this chapter. However, it is not a process that can be adopted suddenly 'out of the blue' – it requires forward planning and an understanding of the whole procedure from the start of the laying season onward, hence this brief comparison with natural incubation. The latter, using either broody hens or the actual mothers, can be effective. However, you will not be able to rely on a specific timescale for the process, nor will you have any real control over the outcome. We therefore consider any results from the hens to be more of a bonus than a certainty.

Using artificial incubation techniques are the generally accepted way of doing things nowadays, especially with pheasants. We followed convention, not just because of our control freak tendencies, but because we wanted a reliable method of raising viable chicks within our own time frame. Our *planned* production numbers are reached solely by using this method.

We will take you through each stage of the process in detail. It may seem overly fussy but please stick with it – it's necessary to attend to some of the apparently inconsequential issues that we now know are important.

Laying Cycles

First, you need to be familiar with the timing and laying cycles of each species. This will give you some idea as to when you need to get your incubators ready for action and cancel all forthcoming holidays!

The pheasant hen will begin laying in March or April. A bout of good weather may trigger the process, although over four years recording the same data, our hen pheasants have shown consistency and started more or less at the same time each year regardless. The results are similar for the partridges, although they begin their cycle later, from late April.

The laying ability of the hen pheasant in captivity is impressive. I mention captivity because, in a natural setting (apart from some exceptions, which I note below), there would be a lessened instinctive need to continue to produce so many eggs. In our experience, if the hen is fit and healthy she can lay around seventy eggs per season. These may vary in shape, size and shell quality and, if some kind of trauma has occurred, the egg may not even have an external hard shell. That said, the common hen pheasant is undoubtedly a prolific layer. The red-leg partridge hen in captivity is much more circumspect and sparing with her energies but, given the right conditions, you can still expect up to fifty eggs per season.

The signs that the hen pheasant is ready to begin nesting are typical. She will start to peck at, and flick strands of straw or dried twigs over her shoulder, and 'nestle' down, making circular movements. But before this happens she will probably lay several eggs in completely random places. There is probably a perfectly logical scientific explanation for this but, to be honest, I do not know what it is, and in any event the outcome is the same.

To try to encourage some order to the aviary laying process you can replicate the ideal nesting site. For the pheasant, this will involve fairly simple structures. She will be naturally drawn to semi-shaded areas, so make sure you have several. However, even if you do manage this successfully, if you start to pick eggs too early she will simply 'up sticks' and revert to laying randomly until she thinks it is safe to try again. Don't be too surprised by this behaviour. In the wild, even without human interference, she will typically make between two and four attempts to create a clutch of eggs. This could be prompted by a number of different factors, including predator destruction of the nest, a bout of extreme weather ruining the eggs or nest, or simply an instinctive dislike of the chosen site. But, once settled, she will lay eggs at a rate of about one per day until her clutch is complete.

It's fair to say that, in an aviary, some hens will never settle, preferring to contribute to other nests. However if a hen does *appear* to be nesting it's preferable to give her a couple of days before dashing into action. When you start, leave a few eggs (which you will need to mark) in the nest before you pick the fresh ones. We have also tried replacing the 'picked eggs' with artificial ones to keep the hen steady but, although a popular practice with some, this has rarely worked for us. Unless she's really focused she is apt to boot them out and abandon the nest overnight. The ideal situation is where the hen forms a nest and sits on her clutch straight away. However, it shouldn't matter if hens take a little longer to establish a nest, or don't manage to at all, so long as they keep laying. The quality of the eggs is not usually compromised – just make sure that all 'lost souls' strewn around the cage are picked very frequently (a minimum of twice each day). This is vital to avoid the risks of damage from extreme weather changes, or absorption of harmful bacteria from a soiled hatch point.

The partridge hen is more particular than the pheasant – we do not have evidence of partridges laying eggs in random spots, like gateways, hard up against the drinkers, or in the middle of the pen like pheasants. They seem to be naturally attracted to a shady, well-protected area, and also produce simple nests on a soft soil base. So to try to accommodate both species you need to have a mixture of straw, twigs and plenty of dry, cosy nooks and hiding places in your breeding pens to stimulate the natural process.

Raised Pens

Another way of managing the laying process is through the use of raised pens. This basically involves raising the nesting area off the ground so the bird is laying on a base of netting. Breeding groups or pairs (depending on the species of bird), are penned in together for the duration of the laying period. The clear advantage is that this method produces a much cleaner egg because the droppings and debris fall through the holes in the wire. The advocates would argue that it is better for the birds because their welfare can be closely monitored in a highly controlled environment. Separated dust and 'recreation' areas are added to the better designed ones.

Much controversy has surrounded this subject and has often been associated with the accusation that birds are forced to live in artificial 'barren', cramped conditions that deprive them of anything like a natural setting. Reported negative outcomes include severe outbreaks of feather pecking. However, it would be unfair to tar all users of this technique with the same brush. The NGO opinion is as follows:

> In essence, we think that good management within any system is more important than the type of system itself and we believe that individual shoots and gamekeepers should be left to decide for themselves which laying/rearing systems to use or to buy from. As long as legislation requires a satisfactory overall welfare outcome – which the 2006 Act and the Government code (Code of Practice, for the Welfare of Gamebirds Reared for Sporting Purposes, effective in England, Scotland and Wales, January 2011) does – the choice of system should be a matter for individuals and not for Government.

Thank goodness for common sense, as always 'on tap' from the NGO. Whilst we do not use raised pens we understand the theoretical benefits. If you find it an attractive alternative, check with the NGO or Defra (Department for Environment, Food and Rural Affairs), to access the detail. The Code gives practical expert advice and is very helpful.

Incubators

Before you begin picking the eggs you need to be organised with your incubation equipment. This is not a piece of kit that can be sourced and plugged in at the last minute. It needs to be fully installed, tested and ready to go two days in advance of your first 'pick'.

Incubators come in many shapes and sizes and conform to two main design types: still air and forced air flow. With still air systems the heat is supplied to the top of the machine. This means there can be a relatively significant variation in temperature between the bottom and top of the unit. The fan 'forced air' system works on the same principle as a fan-assisted oven. The heat is circulated thereby creating a more consistent ambient temperature.

The information that follows clearly describes the differences between the systems and may help with choice. It was supplied by Brinsea UK, a specialist manufacturer of incubators, and is reproduced with their kind permission:

> Still air incubators are generally heated from above the level of the eggs and exhibit marked temperature and humidity gradients between the upper and lower levels so that the tops of the eggs are up to 4 °C (7.2 °F) warmer than the bottoms.
>
> Forced air circulation through the introduction of a fan into the incubator dramatically changes the situation and eliminates the temperature gradient and variations in the relative humidity for all practical purposes. If eggs are to be set on different levels in the same machine it is essential to circulate the air mechanically so that all the eggs are exposed to the same temperature. However, since many of us are concerned with relatively small numbers of eggs which can all be set on one level there is a real choice to be made – forced air or still air.
>
> The adjacent diagram illustrates the kind of variations that may be expected within a still air incubator. It shows that definitive measurement of the relative humidity (RH), as well as temperature, becomes very tricky in still air conditions.

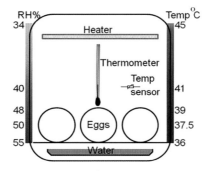

Diagram to show the variations that may occur in a still air incubator.

In practice it is rather inconvenient to measure (or control) temperature or humidity at 'mean' egg temperature that presumably is halfway up the egg.

Thermometers and sensors placed too close to the eggs are vulnerable to damage and contamination and are liable to mis-measurement due to direct contact with eggs or chicks. The metabolic heat of the embryo will raise the egg temperature above that of the air you are trying to control. So control and measurement in still air incubators is normally done above the eggs, requiring some correction to allow for the temperature gradient.

To complicate matters further, the temperature gradient is itself variable, dependent upon outside temperature. In cold conditions it is necessary to raise the temperature slightly at the top of the eggs to achieve the same mean temperature because the egg bottoms are now colder. Even the thermometer may be affected by the temperature gradient. A slight increase in reading will result from the stem being in a warmer zone than the bulb. This is due to thermal conduction taking place down the stem and raising the bulb temperature slightly above its surrounding air.

All these difficulties can be eliminated at a stroke by incorporating a fan. Almost all experiments carried out by research workers studying the process of incubation are conducted in forced draught conditions. This reduces the number of variables and makes for a more predictable environment.

So why bother with still air? Cost? Obviously a fan adds to the cost of a machine. Cheaper machines are therefore usually still air. But still air incubators need to be better insulated than those with fans to keep the temperature gradient within limits. Combining good insulation with a hygienic construction narrows the cost advantage.

All of our incubators are forced air.

SELF-BUILD/DIY

There are a number of websites and articles offering ideas about DIY incubator construction. This obviously reduces costs and is probably OK for a small project. However, we do have concerns about the materials often proposed. Hygiene is a vital factor during the incubation process and materials such as cardboard and wood do not compete well with plastic and metal in that respect.

Number of Incubators

This is obviously determined by the number of eggs being processed. In the season prior to writing, we used six incubators with a total capacity of 420 eggs to process just over a thousand eggs. We also used a seventh with different settings, as the hatching incubator. With our system of daily picking (up to twenty eggs per day) and fertility checking (candling) after twelve days, these ratios worked well.

Temperature and Humidity Measuring Equipment

Commercial thermometers, and particularly hygrometers (humidity measurement gauges), vary in their accuracy, and accuracy is essential. But, rather than spending lots of money equipping each incubator with expensive kit, we have one very accurate reference thermometer and one reference hygrometer. We use these to calibrate the often much cheaper versions supplied with the actual incubators.

Low-cost hygrometer in an incubator.

Most avian species incubate best at a mean temperature of 37–38 °C (98.6–100.4 °F). All contemporary research indicates that temperature control is an absolutely vital factor. As Brinsea notes, in still air machines the temperature reading would depend on the position of the thermometer, which is quite critical – so follow the instructions carefully about adjusting the height. It will also depend on the construction and temperature gradient in the incubator, so again, follow the instructions. If the instructions are unclear, set the thermometer just above the top of the eggs and run the incubator at an indicated temperature of 39–39.5 °C (102.2–103 °F).

Don't forget that no incubator has perfect temperature distribution. Heat losses from the cabinet must be balanced by heat provided by the heater. The process of transferring heat from one to the other necessarily involves a temperature drop – even with a fan – and this drop will mean that some eggs are warmer than others. To keep these differences small, operate the incubator in warm, steady conditions. Ideally use a thermostatic electric convector heater to maintain a steady room temperature of 21–24 °C (70–75 °F) day and night.

Prior to hatching we have all the incubators running at 65–75 per cent RH. But three days before the due date the eggs are transferred to the hatching incubator running at 90 per cent RH. Low air speed and high humidity give the best hatching results. And it follows that, in forced air incubators, the RH needs to be high to prevent excessive drying of exposed membranes. Of course, still air incubators have an advantage in this respect as Brinsea confirms: 'In still air incubators this problem is much less severe and a dramatic rise in humidity accompanies the first birds out, which no doubt helps those that follow. Temperature in hatchers is usually run about 0.5 °C (1 °F) lower than during incubation to compensate for the high metabolic rate of the emerging chicks.'

Many professional game farmers use both types of incubator. Eggs are set in fan air machines for the incubation period and transferred to still air units for two to three days before the due hatch date and left undisturbed.

Buying Your Machine

There are many different sizes of incubation unit on the market, so you have a number of options. If you're anything like us your first choices may not be your last. Ours are a real mixed bag that reflect the evolution of our rearing activities rather than having got the maths wrong in year one.

The basic design is common to each. Eggs are set in rows and separated by dividers which are usually removable. If adjustable, they slot either end into a

'keep'. This allows you to alter the width of the row to fit the species of egg you are incubating. There will also be some form of receptacle to hold water that creates the humidity levels, and a thermometer.

There are two types of turning mechanism available. One is manual operation where, depending on the design features, either a handle or lever enables movement of the egg tray. The other is an automatic version that shifts the tray using a small motor. The manual is cheaper but you do have to remember to turn the eggs at least twice a day. So, this is entirely a question of convenience/cost. We have had good results with both. In fact the best results from our 2011 season were from manual turn incubators. Just factor in the time, and don't forget to rotate the eggs every few hours.

When you make your purchase, as with any electrical device, don't just take it out of the box, plug it in and expect it to work perfectly. Read the instructions and test it thoroughly before use. Assume that it may have a few glitches. Run it for at least two days before you intend to use it, and make sure your incubating room has an even air circulation and ambient temperature of around 22 °C (72 °F).

We completely support expert opinion on humidity levels, but there is a caveat to this. Following the manufacturer's guidelines is, of course, important, but may not always be enough. In practice, as well as checking the reliability of the kit, you need to factor in your actual usage of the machine. For the first two years we could not understand why our hatch rate was relatively poor. We achieved our target release numbers but needed to use more eggs than we felt necessary to reach our goal. We went back to checking out the incubating machines themselves. To our surprise, in spite of following the instructions exactly, particularly in relation to keeping water levels constant, we found in each case the humidity levels were too low.

Pheasant and partridge egg shells are naturally dense (approximately 9 per cent and 10 per cent respectively of the overall egg weight). Therefore a successful hatch requires high levels of humidity otherwise the shell casing develops trauma. As we know, the moment the chick penetrates the outer membrane the inside liquids accelerate their tendency to become viscous and start to dry up very rapidly. The situation is obviously worsened by using a forced air unit. The hatching process is exhausting for the chick anyway, but it can become overly weakened and tire more easily if excess effort is needed. If the humidity levels are wrong, the chick will need to rest 'in process' more frequently, with the outer shell and membrane now pierced. This increases the risk of it becoming trapped, and failing within the rapidly hardening, sticky inner layer. The result can be a needless death to an otherwise healthy chick, and a pointless waste of incubation space.

I mentioned *actual usage* because it may be fair to conclude that some of the problems we have experienced with temperature and humidity levels were our own fault. For example, one of our machines will take sixty eggs. Typically, we set our eggs in batches of approximately twelve at a time. This, quite possibly, would make most manufacturers shudder at the sheer abuse of the machine intended for a one-set only process. But we have to be driven by the practicalities of the process and the need to compromise. This means we need to open the incubator at least five times to set the eggs. Whilst we work quickly, precious temperature and humidity constancy will be affected. However, this is unavoidable. Further disruption is caused by removing eggs to be checked for fertility, and refilling the interior reservoirs, a necessary daily job (easily avoidable if manufacturers considered fitting a tube and external bung). Obviously we try to combine these tasks where possible. To fix the humidity problems, we now accurately monitor RH using our separate calibrated hygrometer, and alter water levels accordingly, adding a second reservoir if necessary to compensate.

Having described this process in detail it is interesting to compare it to the natural setting. If we liken our treatment of the eggs during the mechanised incubation process to the hen sitting on her clutch (and not equipped with RH devices or thermometers), it's fairly clear that our eggs enjoy much more stable conditions. This almost certainly results in a higher yield. It's true that there are many other variables that adversely affect natural hatchability not present in the incubator, but machines do allow more consistent treatment. So, whilst the instructions that come with the machine are of course very important, if your particular way of working differs from the official intended use, be flexible and adapt them sensibly. It should enhance, not diminish, your success rate and will never be as uncertain as the natural clutch setting.

The Incubation Process

Gathering and Setting the Eggs

Once the laying cycle begins you have several choices to make. Do you wait until the hen has created a nest and settled on a clutch? Do you pick the egg the moment it appears and set it in the nearest incubator? Do you gather and store the eggs until you have enough to fill an incubator? We experimented with most of these ideas, with varying degrees of success and failure.

In year one, almost as soon as the pheasants started laying, we rushed in and picked all the eggs. We did not want to wait until the birds started nesting so were happy to take 'all-comers'. Following accepted learning we set only the 'best', most conventional-looking eggs, and discarded the rest as advised. The 'classic' pheasant egg, according to our research, has an acceptable colour range from light tan to pale olive green and is smooth and slightly glossy, with an indicative overall measurement (diameter × height) of 35 × 46mm (approx. 1 × 1¾in) and a weight range of 30–35g (1¹⁄₁₀–1¼oz). For the red-leg partridge the egg colour should be yellowish-white with reddish-buff or greyish markings; it should be smooth and non-glossy, with an overall measurement (diameter × height) of 31 × 41mm (1¼ × 1⅜in), and weight range of 18–21g (approx ⅝–¾oz).

We rapidly filled four incubators and then realised that, since the egg production line is fairly relentless, we had a glut. We read more and decided to store the surplus for the maximum number of recommended days, which averaged out at eight. We filled available spaces made by newly hatched chicks whilst at the same time buying some more incubators. Matters were rapidly getting out of control!

Meanwhile, because it had not occurred to us to mark any of the eggs (not having read that bit of received wisdom), it gradually became clear that we had no reliable idea of the due hatch dates for each batch, and indeed the number in each batch. This inevitably meant that we could not be sure whether or not an egg was overdue, the consequences of which was an egg exploding inside the incubation unit. This disastrous situation meant we had to dispose of all the potentially contaminated eggs in that particular machine, and disinfect the kit thoroughly before putting it to use again. We will not repeat this elementary error and have since recorded each stage of the process!

Basic dos and don'ts

Having demonstrated other similar fine examples of misguided enthusiasm, we offer some basic interim dos and don'ts associated with setting your eggs.

1. You will have a target number of poults you want to release: to be on the safe side assume a high failure rate and set more than you think you need.

2. Don't rush in and pick every egg as it is laid – your hens will be laying for a number of weeks. Bearing in mind the relative labour intensity of the job, decide how long you want, and can afford to be involved in, the incubation process. That will help determine your timing.

3. Establish a collection routine and pick eggs at least twice each day, leaving the final visit as late as possible, and preferably after 6pm. This reduces the numbers that are left out overnight. If your birds are laying on the ground, increase your pick visits during periods of extreme weather to avoid the eggs becoming spoiled.

4. NEVER handle the eggs roughly when you are picking and transporting them. The embryo is extremely sensitive at this stage and poor handling may affect the fertility. The yolk that nourishes the developing embryo is able to rotate to keep the embryo upright, but it won't take kindly to an off-road session or being overly juggled. So use an old egg box and treat it gently.

5. Make sure you wash the egg in a specific avian solution as soon as it is picked. The solution needs to be warmer than the egg to prevent contamination being sucked into the egg via the shell.

6. Make sure you mark every egg that goes in the incubator. Use permanent (non-toxic) marker pens – they don't affect the embryo, and won't wash/rub off. We chart each one. It doesn't matter which method you adopt – just be clear as to how many you have, what the expected hatch date is, and note any relevant idiosyncrasies. The aim is that, at the end of the incubation period and several fortifying gins later, you still understand your own system.

7. Some advocates recommend that if you use a manual turn incubator you should mark each egg with an 'X' on one half and 'Y' on the other. This makes it clear if an egg has become lodged and failed to turn correctly.

Egg selection

When you are selecting eggs for incubation do not be persuaded by the suggestion that all but the 'perfect' egg should be discarded. This is especially important to the small producer, for whom each egg is important. That may seem contradictory given that so many eggs are laid in a season, but if you want to keep to your own timetable then you may have to compromise and take some risks with apparently sub-standard eggs. That was the quandary that caused us to record each pheasant egg set for a complete laying season – whether perfect or not. We were also interested to see whether the reading we had done so far amounted to commentary about aesthetics, rather than scientific proof.

There is much advice in the public domain that states the following eggs are likely to be 'non-starters':

> Underweight
>
> Calcium deposits present on the outer shell (chalky)
>
> Poor shape (generally too round)
>
> Colour abnormalities (green, red, pale)

We found that eggs with idiosyncrasies like these are certainly worth noting, but should not be rejected immediately out of hand. We tested this 'non-perfect egg' theory in the following way:

- Each egg designated for the incubator was weighed twice using two sets of digital kitchen scales. The two readings compensated for any possible inaccuracies caused by using a domestic unit. According to the British Trust for Ornithology (BTO), the perfectly formed pheasant egg averages 31.5g (1¹⁄₁₀oz), and the red-leg partridge egg average is 20.1g (⁷⁄₁₀oz). Any egg that did not conform to the range was marked and recorded. Extremely underweight pheasant eggs (below 26g/⁹⁄₁₀oz) were discarded. We did not have any extreme weight differences in the partridge eggs so did not use them as part of our study.

- Similarly, any egg that was an abnormal colour, unusual in shape, size or texture, had numerous external calcium deposits or any other blemishes/deformities was recorded, but nevertheless still set in the incubator. The ones (very few) that were rejected for incubation were tiny, completely deformed or badly stained with mud residue, even after washing.

- After twelve days in the incubator, the eggs were candled (see later this chapter), to determine whether they were fertile and, if infertile, discarded and recorded.

- Once the fertile eggs had hatched (or not) we recorded the outcome against the original data collected. Successful hatching fell into two categories – *zipped* and *assisted* (very apt terms we came across!). This differentiation is vital, especially if time is at a premium for you. At this stage, however, it is only necessary to point out that both categories resulted in healthy chicks.

- The results on the adjacent charts show relative hatch success rates of eggs with different shape, texture, colour and weight:

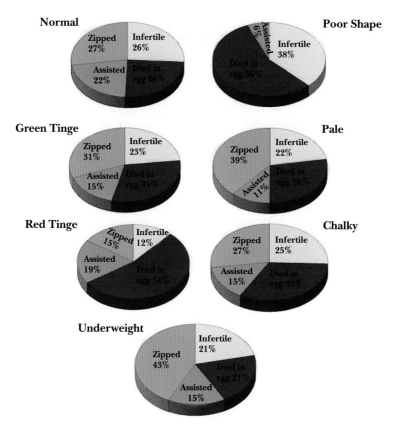

Comparison of hatch success rates.

These showed, within the confines of statistical relevance, that there is one strong conclusion and one slightly weaker conclusion to be drawn from these results.

Strong conclusion – eggs of poor shape produce a very poor yield. The fertility rate was low and none of the fertile eggs hatched by themselves. The overall yield was only 6 per cent, and all had to be assisted during their birth.

Weaker conclusion – eggs with a red tinge gave a poor yield. These showed a good level of fertility but a very poor yield owing to high death-in-egg rates.

A further surprising indication was that the 'underweight' eggs (26–29g/ ⁹⁄₁₀–1oz) had the highest success rate in terms of unassisted births (zipped).

Obviously we need to be aware of statistical significance, and the large number of variables affecting this complicated process. Just one laying season cannot be held as firm proof, which is why we will pursue the conclusions in more detail during following seasons. But, for the moment, we would not

recommend discarding the smaller eggs unless they are grossly underweight (less than 26g/¾oz).

Running alongside this study we tested another piece of advice. This recommended that the first few eggs laid should not be picked for incubation purposes. The reason given is because the embryos are weaker and there is a greater propensity for the chicks to fail. We found no compelling evidence of *embryo weakness*. What we did find was that the *fertility* level from our first eggs was low, but the yield from the fertile eggs was good.

The week by week results from our eight-week picking process are presented on the accompanying chart.

Week	Fertility Rate	Yield From All Eggs	Yield From Fertile Eggs	Comments
1	60%	48%	80%	
2	70%	38%	54%	Storm
3	72%	48%	66%	
4	73%	39%	53%	Fox Attack
5	84%	59%	70%	
6	79%	53%	67%	
7	83%	67%	80%	
8	75%	43%	58%	Storm

Weekly yield.

Throughout our process we were careful to note any exceptional events that might have affected the mating and egg-production process. We had two severe storms, one during week 2, and the other during week 8. Rainfall was not extreme but the winds were gale force. During week 4 there was a fox attack, not directly on the pheasants or partridges, because it could not get into their pens, but it killed one of the free-range cockerels which was only a short distance from the game bird pens. Given the very low yields seen during those particular weeks it is difficult not to conclude that perhaps the need for peace and tranquillity may have a much more serious effect upon yields than factors such as egg weight and colour.

Of course, we do understand the caution expressed by advisers on the subject, especially if their advice is aimed at the mass production market. They want to be as sure as possible of achieving a successful outcome for each egg – but then so do we.

We must now continue to add further research into this aspect so that we are able to broaden the statistics. But our early reaction is that, just because the egg may have a blemish or two, or is slightly underweight, this does not mean it is automatically going to be a dud. If, as we have done, you trawl through all of the contemporary advice, you will see that the acceptable weight range for pheasant egg has a massive 12g (⁴⁄₁₀oz) variation! Perhaps then there is scope for the view that nature is fairly variable and probably there is no such thing as 'the perfect egg'.

Storing Eggs

If you have more eggs than space in your incubators you may need to consider storing some. We put ours in old egg cartons and keep them in a cool-humid storage area. Ideal conditions are around 13 °C (55 °F), and around 75 per cent relative humidity. Advice varies widely on the maximum length of time an egg can be kept before it begins to lose its fertility. This ranges between four and fourteen days for pheasants and longer still for partridges.

We recorded different batch storage time periods, and our results show that, for all eggs stored longer than four days, there is a steadily accelerating decline in the hatch success. On that basis we now use four days as our maximum. If you store eggs for longer make sure you tilt them *gently* from side to side each day to stop the fluids congealing inside. We do this anyway as a precaution. You can manage this effectively by slotting a wedge of wood under one end of the container. Each day move the wedge from one side to the other, allowing the contents of the eggs to shift. The advantage of this method is that you do not need to handle the eggs and risk damaging the protective cuticle. Don't forget that eggs are porous and highly sensitive. The cuticle seals the pores and helps shield the embryo from contamination and reduces moisture loss from the egg. In Nature it eventually dries and flakes off. It can easily be rubbed and washed off, so the least interference the better.

You should also make sure that you store the eggs pointed end down. There is an air sac at the rounded end of the egg, which needs to be kept up to help the egg retain its moisture. This, in turn, keeps the yolk more in the centre of the protective albumen.

Even if you are ready to incubate immediately it is often recommended that each egg is stored for twenty-four hours to allow the albumen to settle, thereby helping to eliminate all risks of internal trauma to the developing embryo. Actually, we do not do this. Our (non-stored) eggs are picked within twenty-four hours of lay. They are measured, cleaned, recorded, marked, left for a short period to bring them to room temperature and then set in the incubator. Our pens are close to the incubating room so there is almost no travel and 'unsettling' disruption to worry about. We have not had any problems resulting from this treatment. So the advice to settle eggs may be much more appropriate for those buying-in eggs that have been in transit for some distance.

When you are cleaning the eggs do remember the functions of the eggshell and cuticle. In addition to the functions just mentioned, the porous nature of the shell allows oxygen, carbon dioxide, and water vapour to pass through, thus allowing the chick to breathe. In order that this process can continue without harm to the chick, make sure that any cleaning solutions you use are 'bespoke', and don't use excessive quantities that wash the cuticle off completely.

Setting the Eggs

Once your eggs have adjusted to room temperature you need to be as quick and efficient as possible. This is especially important when you are involved in a multi-set process using the same incubation machine. Don't forget you want to minimise heat and humidity loss. When setting the eggs, keep each channel filled with similar-sized eggs. This helps them rotate evenly. For example there is no point having a partridge egg in amongst the pheasant eggs and just hoping for the best. It probably won't roll. And, if you don't have enough eggs to completely fill a row, then balance them evenly in the middle of the row to avoid uneven air circulation.

Candling

For the first-timer the incubation process can cause some anxiety. Worrying that you may not have followed the process correctly (and, even if you have, concerns that you have a box full of duds), are amongst the commonly felt fears. There is a proactive practical method of finding out, and this is by 'candling'. It is a non-invasive process that enables you to take a look from the outside 'into' the egg to establish whether or not it is fertile. Executed properly, early detection through candling is extremely effective and can save you time and space, but don't do what we did to begin with and peer at it owlishly with the egg the wrong way up!

After about twelve days in the incubator you should be able to establish accurately whether or not the egg is worth keeping. To candle you need to darken the room, remove the egg from the incubator and hold the round, broader end downwards. Place a torch or bespoke candling device on the base to see whether there is any evidence of life inside. If the egg is fertile you should be able to discern an opaque shape with a very clear air sac at the top of it. Often veins can be seen on the surface of the opaque area. This is the developing embryo.

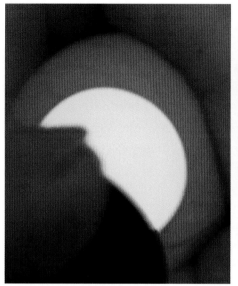

Candled egg – fertile.

Candled egg – infertile.

If, as can be seen from the second of the accompanying photographs, the egg is completely translucent, it is not fertile. To further check this move the egg from side to side whilst candling and you will see the undeveloped yolk sloshing 'untethered' around inside.

Equipment-wise you can certainly use an ordinary torch, but resist the temptation to dig out your ancient 'Pifco'; it's too weak. The beam needs to be small and powerful otherwise you just create a glare rather than a concentration of light that focuses on the interior of the egg. Alternatively, you can buy the real thing: having tried both we genuinely feel that buying a candling torch is a worthwhile investment. They do give much more accurate results. The device is basically a high-output LED (light-emitting diode) torch, with an end shaped so that it can be positioned snugly on the egg to give maximum interior illumination. There is a wide price range but ours weren't expensive. The kit usually contains two or three different-sized end pieces that allow you to candle several different-sized eggs.

During this process you must work quickly and efficiently. Return the egg to the incubator as quickly as possible after you have finished, thereby minimising heat and humidity loss. Remember also to handle the egg gently to avoid any unnecessary trauma to the embryo and cuticle.

What candling won't do is guarantee you a successful hatch. All sorts of things can happen during the intervening period that cause an embryo to fail. However, it does take out the early non-starters.

The Hatch

Three days before the due hatch date we transfer the eggs to the hatching incubator. In these last few days we try to interfere as little as possible and avoid opening it unless absolutely necessary. We plan our batch hatchings to transfer a complete group at the same time.

During this final stage several things can happen and, depending on personal preferences and available time, a decision has to be made whether to assist the hatching process. We categorise the various hatch states as follows:

High-speed 'zipper'. This is the best of all. It occurs when the chick is able, quite naturally, to pierce a hole in the outer shell and literally peck its way out. The chick 'unzips' by using a small horny growth (sometimes called an egg tooth), at the tip of its upper mandible to break through the eggshell. This 'tooth' disappears completely within a few days of hatching. Some chicks will break the shell then pause, then start zipping. High-speed zipping, once started, usually takes less than thirty minutes and requires no intervention.

Tired zipper. Zipping is a very labour-intensive process for the hatchling that is using its beak as a pneumatic drill. This sometimes means it needs two or three attempts to fully zip and a rest in between. If you are keeping a watchful eye on the process through the incubator observation window, try not to overreact and rush in to assist if the zipping has stopped. The chick may just be having a kip.

Stalled zipper. If, after an hour, a chick has partially zipped and stopped it may be so tired that it can't continue and will die in the egg. If this chick is to survive it needs help.

Peeking. In this situation the chick successfully breaks a small hole in the egg but fails to start zipping. *However*, do not start assisting a peeking chick until you are certain it is not going to proceed to a zip. Some chicks peek prematurely and, since they are not yet fully developed, they will die if interfered with too early. At least one day after due date should elapse before intervention and, even then, this should be done very cautiously. Often chicks tell you they want help by 'pipping' when they sense local noise or movement.

Wrong-enders. This is a situation whereby the embryo is misaligned in the egg and, whilst peeking can occur reasonably easily, the chick is in the wrong position to start zipping. These always need help.

Premature death in the egg. This occurs in a disappointingly high proportion of fertile eggs. Whilst siblings are peeking and zipping, these eggs do nothing and, two days after the due hatch date candling will show that, unlike a chick that is developing slower than the rest, there is no movement within the egg. In this case, for reasons of hygiene, it should be immediately discarded.

Assisted Hatching

You will see from the above that there are often situations where you may have to assist in the birth. During our last season of hatchings we successfully assisted 201 of the pheasant chicks, significantly improving our overall yield.

This process is not exclusively associated with an overdue egg. Some chicks struggle during the normal, or even premature, hatch period. The typical signs will be an audible 'pipping' sound when the egg is held to your ear at due hatch date, but there is either no visible piercing of the outer shell or a just a small crack. At this stage leave well alone to see if Nature can find a way for the chick to break out. On your next visit (after a couple of hours or so), if the pipping has become less frequent or fainter, either of these indicate that the chick is weakening with the effort. Action may then be needed. You may also find that the chick has managed to make a hole but has apparently got its beak stuck partway through the outer casing (see photograph). This is not so much caused by opening the incubator lid too often and allowing humidity levels to escape; rather these situations can often occur after the due hatch date has passed, suggesting that the chick has struggled for some time, and is now rapidly becoming exhausted.

This is a dangerous situation because the embryonic fluids quickly congeal and harden around the chick's beak thereby compromising its ability to breathe. Chicks can suffocate in this way. Colour danger signs become evident, with the membrane around the beak often taking on a yellowy/ochre hue.

Much is said about the pros and cons of assisting chicks to hatch. Some favour the Darwinian concept of survival of the fittest, claiming that by assisting chicks that would otherwise die during hatching you are perpetuating a weaker strain. It is certainly true that we have found chicks in eggs that have been discarded by their mothers that look to have perished at a very similar stage. It's possible that the mothers have an instinctive awareness that one chick is weaker

'Peeker'.

than the other, and so focus their attention on the strong ones. Or, it may just be a case of poor incubation.

Our position is that we will not leave chicks to die if they still have a chance of survival. Our method of working in this way has resulted in hundreds of healthy and contented birds that would otherwise have been destined for the bin. It is, however, a fact that assistance is a time-consuming process, so it becomes a personal choice as to whether your situation allows that.

Equipment for assisting hatching

If you do decide to assist, you'll need the following:

1. Kettle for sterilising equipment and warming kitchen roll/towels
2. Cotton wool buds
3. Blunt tweezers (ladies' eyebrow types are good)
4. Kitchen roll
5. Wooden cocktail sticks
6. Small pair of scissors with sharp blades
7. Bowls for water
8. Antiseptic
9. Reading glasses, if needed – this is a close delicate process

'Zippers'.

Technique

Working on a clean, sterilised work surface, remove the egg from the incubator and cradle it with a warm, moist sheet of kitchen roll or similar material. Using a cocktail stick or other pointed instrument, begin to flick away small fragments of the outer shell. This reveals the whitish membrane beneath. You should start at the point where the crack has appeared or where the beak has become impacted and create a clean break in the outer shell and inner membrane as you work your way around.

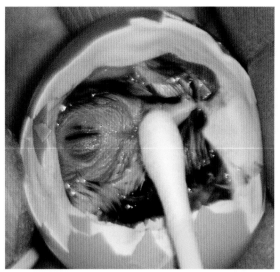

Assisted birth stage 1.

Observation of the classic 'zipper' will show you that it starts midway up from the base of the rounded end of the egg and, working its way around, pecks out an almost perfect circle. This is what you are trying to imitate with your cocktail stick and tweezers.

If, at any stage during this process you see signs of blood, *stop*. This means that, whilst the chick may well be quite physically active, it has started its process too early and must be given a further period of in-egg development. Make sure the area around the chick's beak and nares (nostrils) is clear, and return the egg to the incubator for a minimum of six hours before checking further.

Once an egg is zipped there is a temptation to help the chick out. *Do not do this*. Some chicks will flop out anyway, but others will not as they are still connected by a remnant of the attachment from the yolk to the abdomen. Return the chick *carefully* to the incubator with the egg still in place, and allow the chick to push itself out at its own pace.

Once hatched, it will continue to live off the absorbed goodness from the albumen for three days, and it should be left alone to dry off and rest in the incubator for at least twelve hours. In fact you should not give the chick any

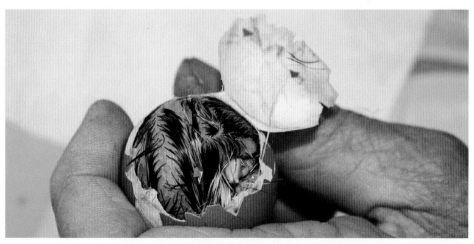

Assisted birth stage 2.

Assisted birth stage 3.

food for at least the first twenty-four hours because this may affect its delicate digestive system.

In the cases of difficult or prolonged births there is a tendency for the outer membrane to become dry and hard. It takes on a light brown tinge. This is where cotton buds, warm sterilised water and tweezers have to be used to soften it off and remove it. Always ensure the beak area is clear when doing this.

The assistance process may take one attempt or several – it just depends on the state of the egg and the chick inside (and also your time and fortitude). You should not attempt it until you are absolutely sure the chick needs help. Having made that decision, if you persevere with patience and care, you are likely to end up with a healthy, robust chick.

Egg with brown tinge.

Partridge chick.

It's probably fair to say that, in the professional game farm, this kind of practice will never occur. They don't have the time. But if every chick counts within your own time frame, and knowing there's an otherwise healthy chick that's just got a bit stuck and needs a helping hand, it can pay dividends. Just cancel the dinner dates for a few weeks – the results are worth it.

Premature or Overdue Chicks

There will be occasions when, despite your meticulous recording procedure, Nature ruins your system, and the eggs fail to hatch exactly on time. This can be very annoying for the classic control freak. If the chick hatches early this is usually fine, especially if the egg has already been transferred to the 'hatching' incubator. However, it doesn't always happen so conveniently. If this isn't the case and you are using auto-turning devices, you need to develop the habit of regularly checking the interior in the run-up to each batch's expected hatch date to make sure you don't have a chick hurdling the partitions inside. Most incubators have a transparent viewing window, which is handy and avoids having to open up to look inside.

If, however, the egg appears to be late, you have a further decision to make. Remembering that the gestation period for a pheasant is twenty-four to twenty-five days and for a partridge twenty-three to twenty-four days, you will need to add some level of flexibility because, unless you witnessed the laying of the egg, you can't be sure exactly when it was laid before you picked it. Generally hens will lay either early in the morning or late afternoon, so it is useful to time your picking sorties with these periods. But, it is still hard to be certain.

The adjacent chart shows that hatch dates do vary. Therefore, assuming you

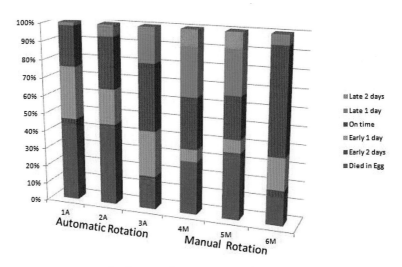

Incubator performance comparison.

have candled the egg to confirm fertility, you should be prepared to leave it unhatched in the incubator for three days longer than the estimated due hatch date. Leave it too long (as we did during our first season), and all hell can break loose, leaving you with another exploding recipe for disaster. So, in spite of an earlier positive candling result, there can be late gestation problems causing the embryo to fail. Don't be sentimental and feel that you're not giving a late developer the chance to hatch. Dispose of it after the calculated maximum 'safe' period. Failure to do this risks having to clear out a whole incubator full of otherwise fertile eggs.

Alternative Use of the Incubating Machinery

Some may consider this idea to be taking economies of scale to an extreme, but we've never liked waste. We started our chicken flock by being given a motley assortment of different-sized eggs. Not a terribly technical beginning, but they were free. The donors didn't seem to have much of a clue about the ancestry (of the eggs that is), just that they were medium-sized hybrids. We popped them in an incubator and twenty-and-a-half days later they all hatched perfectly.

Since then they have been incredibly easy to rear and tend compared to the game birds. Rather a shame really, that we couldn't use them for sporting purposes. We later learnt there are some scientific reasons for this ease in hatching. The chicken has been selectively bred to produce larger and larger eggs. The 'modern' egg of a domesticated chicken is two or three times the size of the egg of its ancestor, the red junglefowl. The big drawback to this though is that the shell material has not increased, so the same amount of shell is stretched thinly over a much larger surface, making it relatively much thinner and more fragile. But, it's great for hatching.

We used our first group of chickens in the following ways, as egg producers, broodies, flock guardians, and for the pot.

The egg producers provide us and neighbours with our daily 'vittles' (giving them to the neighbours always comes in handy when a minor favour is needed). For the broodies we selected the smaller ones that were suitable for incubating rather than crushing game bird eggs, i.e. as close to bantam size as possible.

In order to continue the strain/size, we run a medium-sized cockerel with them so we have a stock of fertile eggs to produce future surrogate mums. We then have the guardians. These are any spare well-tempered cockerels deemed

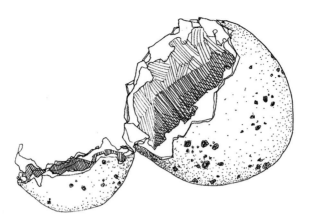

fit and alert enough to act as warning devices against predators. We currently have six of these that roam around the rearing pen compounds, and within the release pen.

The truly surplus birds get eaten. These are generally cockerels with bad attitudes that attack all-comers so serve no other practical purpose than the pot. Perhaps a little of the smallholder instinct is creeping in here and it's definitely not technical, but we're happy with it. After all it's a shame to have your expensive incubating equipment lying dormant when it could be used profitably outside the game bird reproduction season.

3

THE FIRST GROWTH PERIOD

'AFTER ALL, THEY'RE JUST KIDS'

Once the chick is out of the incubator the clock is literally ticking in terms of managing its growth in a methodical way. Your job is to provide sufficient heat, light, water, food and medication during the rearing phase to allow it to develop healthily at a normal rate. You need to follow the simple routine of:

1. Fresh, clean food and water at all times.
2. Sanitary, light-controlled environment.
3. Warm, draught-free living conditions.
4. Gradual exposure to the elements.

If you do so, the chicks should feather-up nicely, and be on schedule for transfer to the release pens from six-plus weeks for pheasants, and ten-plus weeks for the partridges. It is pretty straightforward, but if you are raising birds in batches you need to be organised. So – get a system and chart it!

Once the chicks are ready to leave the incubation suite they will be moved as follows:

1. First, to a brooder hut with a 'nursery' built inside which is temperature- and light-controlled.
2. After approximately three weeks, a run is attached that allows the chicks to experience a taste of outdoor weather and begin very gently 'hardening off'.
3. After a further two to three weeks they will be moved to a separate rearing pen which has several weather-proof shelters. They will live here until they are old enough to be moved to the release pens. ***N.B.*** *These date figures are based on pheasant chicks; for partridges add on two weeks as a general rule of thumb.*

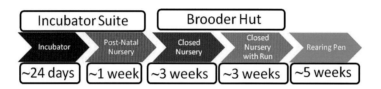

From the incubator to the rearing pen.

The Incubation Suite

Having experimented over the years we now follow a set pattern for managing the first growth period of the chicks. It's likely to be too complicated for some, and impractical for those of you who are raising numbers in excess of 200 in one batch, but the following process will guarantee you a higher yield.

The healthy chicks normally rest in the 'hatching' incubator for two days before being moved to the 'post-natal nursery' (either rabbit or game bird cage). Precise timing depends on the health of the individual. If a chick is weak we will leave it little longer. We also occasionally return them if they begin to fail. This level of observation requires attention to each chick.

The Post-natal Nursery

Pheasant chicks will remain here for up to seven days, and partridges for around two weeks. We monitor them closely, and make sure they are developing properly. The two fundamental cage designs differ. The rabbit cages we use are the plastic and wire varieties. Each has a plastic base approximately 15cm (6in) deep and measures 20 × 50cm (8 × 20in). One design has a concave Perspex lid with a central plastic-coated metal grille overhead, with a latch door. The other has a similar base but with a plastic-coated wire upper section. Both are equally effective.

The bespoke game bird/chicken cages are fully made of galvanised steel. They have a simple wire grille base with a slide-out metal tray that sits on a shelf beneath. One-third of the cage is enclosed by exterior metal walls, and the other areas are enclosed by rigid mesh.

In all cases we fit an infra-red heat lamp. We developed a handy design in which the bulb connection is mounted on a 50cm (20in) long hollow tubular steel rod. This allows us to raise or lower the heat lamp according to the chicks' needs. The lamp kit sits on a wooden block and has grooves to prevent slippage on the metal grille.

Malformations

The bases of the rabbit cages are plastic and need modification. This is extremely important. Pheasants in particular have a tendency to develop one of two lower-leg malformations:

Splay leg. This usually occurs when the bird doesn't have enough friction under its leg when it stands (so it can't 'push' itself up or hold itself up straight). The legs grow out sideways and make the bird looks as though it has performed 'the splits'. The cause of this condition is often a floor surface which is too slippery for the bird to grasp well. Such a surface can also be present in a broody nest, so make sure the nest base is not slippery.

There is also a medical explanation for the condition which is associated with a diet that is either too low in calcium, or too high in phosphorus. Either discrepancy may aggravate or precipitate this condition.

If you're going to tackle splay leg it's best to do so in the first few days because the older the bird, the more calcified the bones become, and therefore more difficult to correct.

Crinkle toe. This is a mild bony malfunction that chicks can either be born with, or can develop if the base covering is wrong.

If faced with a malformed hatchling you need to decide whether to let Nature take its course, destroy it, or try to repair it. For a professional game farmer there is probably no decision to be made – the bird will simply be despatched quickly. Why? Because attempting to repair a bony deformity takes a disproportionate amount of time and patience, with no guarantee of success. Nevertheless, we have tried to do so on many occasions with varying degrees of success.

We have also nurtured disabled chicks to adulthood, but this is rarely effective in the long run because the bird soils itself too often. In this case, where our sentimentality dominated practicality, we honestly did the bird no favours.

One of the techniques we use for treating malformed limbs is to support the legs and try to train them back into a normal position whilst the bones are still malleable. This will normally be at two or so days old. It can be successfully achieved using splints made from lollipop sticks (or similar), held together by elastic bandages or by strapping the legs together gently and positioning them in a section of cardboard kitchen roll tube. We have also tried using a 2cm (¾in) Elastoplast which makes an excellent, soft prosthesis. Other materials like a pipe-cleaner can be used, but the 'pad' in the middle of the plaster is a good thickness to keep the chick's legs the correct distance apart. Cut the plaster in half lengthwise to get the correct width and strap the legs around the tarsometatarsus (bone above the foot). This can work well, but takes time and the outcome depends on the severity of the deformity.

Plaster support.

If you are worried about the splinting technique, talk to your vet. That's what we did. The bird needs slow, gentle handling, regular dressing changes (because it may soil itself) and you need to make sure it can eat and drink. If it can't, fill a syringe or similar with tepid water and pop drops into the side of its beak. Adding a *touch* of sugar will give it an additional glucose energy boost and encourage it to drink. Then, holding a small container of food close to its beak, tap the base to encourage it to feed.

In our experience if, after five days, the chick is not progressing, it's likely to be a lost cause. But this procedure *can* work, so if you want to try it, make sure you have the time and commitment to do the job properly.

The Nursery Base

As mentioned earlier (assuming the cause is not medical or genetic), the base covering of the nursery is of critical importance in avoiding these malformations. You will read lots of advice recommending the use of various different materials. These will include shredded newspapers, sawdust, wood chippings and straw. We have probably tried them all and, for the rabbit cages, we now use exclusively exotic bird cage base liner sheets, the type normally used for canaries, etc. If you are not familiar with this material, it looks just like conventional sandpaper and indeed has a base covering of sand. But it is finer, and differs in composition, so avoid rushing to the conclusion that conventional stuff from your local DIY store will do – it won't. It may be too harsh for the hatchlings' feet and, bearing in mind that chicks are inveterate 'peckers', they may be at risk of ingesting harmful chemicals from a non-bespoke product.

Bird liners contain a mix of natural minerals (and sometimes aniseed) and finely crushed oyster shells. These materials combine to help build sufficient calcium and mineral levels that the birds need, as well as a great stable base. They can also help keep the claw length in check and potentially the beak too. This surface is undoubtedly excellent for correcting minor bony deformations, and gives the birds good 'purchase' which helps strengthen the cartilage, essential when walking.

If you decide to use the rabbit cage the downsides are that the chicks will soil the base very rapidly so it will need changing frequently. This is time-consuming and more expensive, although we now buy liners in bulk ahead of the season, which helps keep costs down. Changing them may be tedious but now we only have isolated problems with hatchlings that are unable to correct an early bony defect by walking on this base material. This reinforces our theory that the early incidents of this type were more a consequence of poor choice of base covering than a genetic deficiency.

Post-natal nursery cages.

The post-natal nursery game bird cages with the wire base are a different design from the rabbit cages. They are equally efficient and certainly much more hygienic. This is because the droppings simply fall through the mesh holes to the tray below, which we line with brown paper and dispose of daily. They are designed to be used for rearing birds as small as quail but we do not use them for our partridge chicks (a bit bigger than a quail chick), because we think the base spaces are too big for their feet. This design lacks the opportunity for the chicks to peck at a mineral-based sandy liner, but we adapt by fitting a piece into the base of a plastic plant pot holder.

Introducing Food

When you transfer the chicks to the starter units you need to introduce them to food. Give them bespoke game bird starter crumbs. Don't be persuaded by other similar products, e.g. chicken chick food which is often much cheaper; it won't do. The protein and vitamin content in the game bird mix is much higher and they need it – especially the partridges.

Very sadly we suffered several losses during our early years raising partridges despite following every instruction we could find. Our hygiene standards were high, the food was the best we could buy, and the pheasant chicks were thriving, so what were we doing wrong? We eventually went back to trying to understand how and why the process succeeds in Nature. It's really very simple. The parents will forage and feed grubs and insects to the chicks immediately after hatching. The youngsters thrive off this and learn to search in a similar way themselves, grubs being the key!

In an effort to replicate this behaviour we researched some more advice that suggested feeding the chicks ant eggs. But enthusiasm with the trowel soon wanes when there are more than fifty beaks to feed four times a day for a couple of weeks until they can cope without this supplement, and exist solely on conventional game bird mix. So, to cut down on this endless task we decided to experiment with maggots. We've never looked back; maggots are definitely the thing.

We now supplement the diet of all our chicks with a small supply of maggots twice daily during this first crucial growth period. Packed with protein, they provide an excellent source of natural nourishment and cause something of a feeding frenzy in the pen! We now also use them for all our penned birds during times of stress, e.g. disease outbreak or injury. Maggots are unquestionably one of Nature's nutritional success stories.

We purchase our stock from the local fishing tackle shop and fortunately our needs coincide with the summer fishing season, so there is always an endless supply of different sizes. We use the 'pinkies' (larvae of the greenbottle fly) for the partridge chicks because they are smaller and easier to eat. (Be aware that there is a difference between these small, genuine 'pinkies' and the standard, larger maggots that are sometimes dyed various colours, including a light red. There have, in the past, been human health concerns about some of the dyes used to colour maggots so, to be on the safe side, if you do buy the larger maggots, stick to white ones – the pheasant chicks are fine with these.)

The maggots, which we buy by the pint, are stored in the fridge, which keeps them fairly inert. Breakouts can occur so to counter these embarrassing fridge moments, we place the cartons in Tupperware boxes which safely contain even the friskiest ones. In these conditions they are usable for a week but just make sure you have some tiny air holes in the lid.

Preparation and Routine

Before the chicks are moved to the starter units, make sure their new home is ready. Don't leave anything to chance. Check that the equipment is working and that the interior is constantly heated to an ambient temperature of around 32 °C (90 °F).

In the cages the chicks develop a fairly routine pattern. They will immediately search for the source of heat. We use either an infra-red or ceramic 'reptile' bulb. Typically they will bundle underneath it in a little downy lump until they are warmed up. They will then spread out, forming a circle around the perimeter of the heat rays. This behaviour is important for you to observe. If the chicks remain in a heap in the middle you are probably giving them insufficient heat, so lower your lamp, thereby increasing the heat by a few degrees. Conversely, if they are spread in each extremity of the cage then it's too hot. What you're looking for is the classic 'ring doughnut' shape that is roughly spaced around the lamp beam.

Being babies they will nap frequently, wake then play, eat and drink – then repeat the cycle. When sleepy the partridges will often simply collapse and lie flat, giving a realistic impression of being a corpse, whilst the pheasant chicks are often a little more compact, gently nodding off in the warmth. Fresh food and water need to be constantly available but try to keep these as far away from the heat source as possible. Direct exposure of the food to infra-red light can result in its deterioration (also known as photodegradation). This exposure can cause distinct flavour changes as well as the loss of added vitamin A, riboflavin and vitamin C. If this vital vitamin source is destroyed, the goodness is lost and health problems can occur.

The chicks remain in the post-natal nurseries until they're ready to be moved to the nurseries (brooders). This readiness will be obvious because they will be eating and drinking freely, and will have begun to 'ping' about inside trying to flutter out.

The only real drawback we have found with the design of the game bird cages is the loose hinged 'lid'. It is very wide, and opens fully to a vertical position. Despite our best efforts to lift it only a little when changing food and water, it's easy to inadvertently create mayhem. You may well have the occasional 'Houdini' who manages to squeeze past your hand and pop out. But don't be alarmed; they are usually fairly easy to 'field'. Once out they normally perch on the nearest point and freeze, trying to work out what to do next. If this happens don't dither, just quickly scoop them up and pop them back in. A good pair of cricketing hands is of undoubted use for this kind of fielding exercise!

The Brooders

Currently we have four nurseries designed to contain between fifty and a hundred chicks at any one time. We allow for around forty chicks per square metre for the first three weeks. The brooders are built inside our converted dog kennels, which allows us control over heat and light. In adapting the kennel-brooder houses we lined the external mesh walls with thick black plastic sheeting. It keeps draughts off the nurseries inside; it's easy to clean and gives us complete control over the amount of light we allow the chicks. If they have too much brightness they can develop unwanted behaviour. The favourite is feather-pecking which, once learnt, is difficult to stop. Pheasant chicks in particular can be merciless and quickly injure or kill another bird. So it is very important to strike a proper balance between giving them just enough light to flourish, whilst keeping over-energetic behaviours subdued.

When calculating the size of the nursery always add a little more space than you think you need; this allows for future expansion and can only help the chicks, particularly since they develop so rapidly. And don't forget to factor in space for both food and drinkers, to be positioned a safe distance away from the light/heat source.

Construction

In constructing the nurseries we used existing resources where possible, whilst trying to follow the various best practice rules. For example our walls are made from old doors and sections of marine plywood. You can use a wide variety of

Corners softened with 45° vertical strips.

materials – just make sure they do not contain any toxic coverings. They also need to be sufficiently robust to be able to withstand the rigours of numerous pecking beaks and afterwards being cleaned down with a jet hosepipe and disinfectants.

We started with a basic rectangle. We 'softened' each of the corners with 45° vertical strips. To create an even, safe airflow we drilled large ventilation holes around the top third of the walls. This allows sufficient air circulation but none that is directed onto the chicks. Each nursery has a hinged lid with a very large wire netting observation window. This keeps them safely inside the nursery and helps maintain control over temperature and light. The heat lamp system is fitted within the lid frame. It is designed in exactly the same way as the post-natal nurseries so the lamp can be raised or lowered as necessary.

As ever, remember to check your kit and pre-heat the brooder well before the chicks go in. Put a thermometer on the floor midway between the centre of the heat lamp and the walls to check the temperature. The theory is that the temperature should start at around 30 °C (86 °F), reducing gradually to ambient over a four-week period. This allows the chicks to harden off gently and feather up nicely. Do beware of cold snaps however, and reinstate the heating if necessary.

As for the floor covering we have two types. One is a roughened marine-ply base and the other is concrete, which we definitely favour. It doesn't present a temperature problem; it's easy to clean and is a great surface for the chicks to walk on without slipping. We cover it with a fine sprinkling of aviary sand and a couple of handfuls of untreated wood shavings. Other suggested materials include shredded paper, chopped straw and paper kitchen roll. One word of caution: whichever material you decide to use, do so sparingly. The chicks will instinctively peck at anything and can easily clog up their throats with soggy bits of wood chips, which we have occasionally had to 'oik' out of a compacted throat. The other problem is that most of these materials, when kicked about by unruly youngsters, can quickly gum up the water container bases. So whilst is can be a good idea to provide some additional floor covering, don't have them knee-deep in the stuff. The rest is wasted expense and causes extra work.

Using our system the height of the nursery needs to be at least 1m (3.3ft) – low enough to reach into, but sufficiently high to avoid escapees when you open the lid, and to allow an adequate heat lamp height.

Equipment

There is a substantial range of feeders and drinkers readily available on the market. We have tried several, and experimented with other non-bespoke odds and sods over the years. This has led us to our now fairly impressive motley assortment that caters for each age group.

Drinkers

For the very young chicks in the post-natal nurseries we use chunky plastic exotic bird drinkers. The chicks are fascinated by the air bubbles that rise up inside the tube, and peck at them endlessly. This behaviour is of added benefit in getting any reluctant first-time drinkers going. We slot them in a home-made wooden block so they can't be knocked over, and make sure they are positioned near, but not in, the corners. The drinkers are too small for the chicks to drown in, and it's generally quite tricky for them to defecate in – but it has been known. Another idea is to use a dish. They're cheap and handy but tend to get soiled quickly because the chicks wade around in them. You must also remember to put pebbles or marbles in the base to prevent the chicks from falling over and drowning.

We have also experimented with nipple drinkers, which are very popular with game farms. These drinkers are often supplied with the rabbit cages and are worth a go. However, we found the problem with these was that once the

Small bird 'canary' drinker.

chicks got the hang of pecking at the nipple it became a great game. After a concerted group effort they quickly ended up wading around in soggy base material looking really thirsty. So if you do decide to use them try to get the design that has a mini-trough beneath. We still prefer the adapted canary drinkers.

Once in the nurseries and runs (discussed later), the chicks progress to larger water containers. We start with the fount manual fill drinkers. Ours are the red base, white plastic dome variety. The white dome is split from the base trough, filled with water (various sizes available), and then clipped back onto the base. The whole unit is turned back over and is ready for use. Depending on the design of your brooder you can also use units that are suspended off the floor. These have obvious hygiene benefits but be careful that chicks do not get stuck underneath.

As the chicks get a little older we introduce mini galvanised-steel buckets which lie on their side, resting on an integral lip. These are particularly useful in the runs, and have the added advantage of being a source of great amusement for the youngsters. They will peck at their own distorted reflections, and spend hours hopping on and off the top. We also have the upturned old-fashioned glass milk bottle types that empty into a series of holes and a receptacle beneath. The drinker comes in two sections, using the same system as the fount drinkers. A variety of sizes and bottles are available. If you take this option make sure you change the bottle regularly or clean it with a suitable sterilising tablet. It is difficult to get at the inside walls so poor hygiene risks a scummy bacteria-infested deposit developing.

Occasionally a chick can get wet, often because it has been trampled on by the others as it plays around with the water. This is disastrous for the birds who are not at all waterproofed at this stage, so cannot tolerate being damp. Until they start to develop their first plumage they are highly vulnerable to draughts, cold and wet so try to take as many precautions as you can. Also allow space between the drinker and the wall behind to avoid chicks getting trapped.

Feeders

We adopted a similar trial process for the feeders. In the starter units we now use either plastic plant pot bases, or exotic bird seed containers. The larger troughs are unsuitable for this age group. With the plant pot bases we place a small one filled with food inside a larger one. The birds can access the food easily and the larger base acts as a back-stop to minimise wastage as the birds scratch and forage, and generally chuck perfectly good food away. The added advantage is that the pecking sound is quite audible on plastic, and attracts the reluctant first-time diners. Alternatively the plastic troughs can be used. They clip onto the wire walls or slot into a homemade wooden block, the downside being that the lip is often used as a perch. Spillages and soiling occurs with both but, on the whole, they are effective for the chicks. Both options are cheap, and easy to clean.

Once in the nurseries we switch to using mini-troughs. The plastic variety has detachable bars, which is helpful. It introduces the youngsters to eating out of a larger receptacle, allowing them to access the food without being restrained by the bars until such time as they can reach in properly. The purpose of the bars is to act as a deterrent, preventing the chicks soiling the food. Sadly they are pretty determined in their dedication to ruining an otherwise good meal, and can mistake bars for perches, so once again your vigilance is needed.

Heat and light

These are extremely important: it is imperative that chicks do not get chilled. If this does happen don't be hoodwinked into thinking that placing a cold bird back under the heat will necessarily revive it. Try by all means, but sadly, in spite of appearing to have recovered, a high proportion develop gastro-intestinal problems or liver and kidney failure, and will probably die a few days later.

There are several different types of artificial heater on the market. The heat source is either gas (a favourite with many game farms), or electricity. Gas heaters need to be linked with the correct regulators and ideally have a thermostatic controller to keep an even brooding temperature above the chicks. They are an attractive option if you experience regular power cuts.

We use infra-red electric brooder lamps. They are easy to use and are a popular choice for keepers rearing smaller numbers of chicks. The fitting is usually suspended from the ceiling and comes with various different attachments, the most usual being equipped with a cowling around the bulb that reflects the heat onto the brooding chicks. A guard surrounding the bulb may also be fitted. This is useful if your nursery is shallow and helps prevent injury. With this system you can link several bulbs together, allowing you the flexibility to add and remove individual bulbs to suit the temperature and light needs of the nursery.

Heat bulbs come in different wattages, and are coloured red, white and sometimes clear (often associated with poultry farming). Some bespoke bulbs are also shatter-proof. Some advice you will read even suggests using normal light bulbs. Don't. Our experience says white light over-stimulates the birds, so we use the infra-red type, and also ceramic reptile bulbs which give off no light at all. The ambient or room lighting can then be used for the daytime/night-time cycle.

Another design available is roughly shaped as a table top with adjustable legs. The chicks stand underneath it to access the heat source. It is made from tough plastic materials and is easy to clean. It contains a radiant heater that runs from twelve volts and is rated at fifty watts. Advocates believe these are more efficient than infra-red lamps.

Toys

A further consideration is the use of accessories to keep the chicks amused/distracted. This may sound over-indulgent but at this stage you should try anything to avoid an outbreak of malevolent behaviour. We have a variety of low-cost 'toys' aimed at diverting the youngsters. Some are suspended from the lid, whilst others are stuck to the walls. For example in the nurseries we glue 10 × 20cm (4 × 8in) mirrors on opposite walls, giving an impression of infinity. The birds love their own reflections and of course spend a great deal of time pecking at them. Other odds and sods range from an ageing vanity mirror, entirely redundant for humans during the rearing period, to exotic cage bird rotating mirrors, 'disco' spheres and bells. All completely ridiculous but these mini-diversions allow the chicks to work off a bit of youthful energy whilst causing no harm to one another.

Early Feeding and Health Issues

Once in the brooders we feed, change water and generally monitor the birds at least twice daily. They will quickly get used to a feeding routine so try to keep to approximately the same times. Food for the birds changes in size and ingredients as they develop as follows.

> **Starter crumble** mixed with a sprinkle of maggots: birth to four weeks.
>
> **Second age granules:** four to six weeks.
>
> **Adult pellets** mixed with wheat/cereals: six to eight weeks.

These timings will depend on the type of food you are using, so check with your supplier. When you change the food type, do so gradually by making a mix of both varieties, phasing out the first one gently. Failure to do so may cause the birds to have bouts of diarrhoea, which you need to avoid.

During this first crucial growth period you should also give thought to supplementing the chicks' normal diet with anti-parasite treatment. There are two distinct schools of thought on this subject. One follows the 'if it ain't broke don't fix it' mantra. This implies that if the husbandry practices are followed properly, there should be no reason to supplement their diet with chemicals. The other view is that where you have intensively reared creatures, however high the standard of husbandry, the presence of negative bacteria and parasites is inevitable and a preventive treatment programme is therefore appropriate. For example, one of the most deadly diseases common to intensively reared birds is caused by coccidiosis. This is a very common parasitic disease of the intestinal tract caused by microscopic organisms called *Coccidia*. It is extremely prevalent in aviary breeding and can cause high mortality.

After some agonising, we came up with a compromise. With our vet's help we set up a treatment programme of herbal and chemical products designed to prevent disease and provide a health boost to the chicks. We follow this rigorously and so far have not experienced any disease outbreaks in the nurseries. The full programme is as follows:

1. Seven-day anti-coccidiosis treatment starts end of second week.
2. All-round wormer single dosage given at eight weeks then every month thereafter.
3. Complementary medicine given initially at six weeks then every month thereafter.

The most successful non-chemical product for us is 'Verm-X'. It is a 100 per cent herbal formulation for the control of internal parasites. We use it for all our birds (including the chickens), on a monthly basis year round. The product is intended to provide a natural resistance to parasite build-up, so we think it makes sense to keep the treatment regular. I cannot say that it has totally eliminated parasite outbreaks in the game birds but we think it has reduced the frequency. Conversely, our chickens have never had a chemical in their lives, just Verm-X, and are all completely healthy. So we remain totally committed to the product.

Given the correct treatment, the chicks will develop very quickly in the nurseries. We always try to harden them off as quickly as possible because it gives them longer to acclimatise and learn some survival skills before the shooting season opens. But if the weather holds us back we will respond accordingly.

The Runs

After about three weeks in a completely closed nursery, and always weather-dependent, we fit runs to the nurseries to give the chicks exposure to outside sunny temperatures. In reading this description, bear in mind our layout, which has the nurseries built inside ex-kennels, each of which opens out to an old gravelled quadrangle. We therefore could not follow the standard procedure of allowing the chicks a grassy 'sun parlour' directly adjacent to a rearing field. Instead we put down concrete slabs and fitted 'mini-runs' to them. These were made from extendable wooden garden trellis pieces that we positioned on their sides rather than in the intended vertical form, and lined with sturdy 10mm (⅜ in) square-hole plastic netting. We made covers from wooden slats fixed together by lengths of cable.

These runs have proved to be excellent for the chicks. We just hook them up to a trapdoor in the nurseries, and when weather allows let the chicks out to enjoy the sun. We do this as early as possible, making sure they are herded back in again when the temperature drops. Simultaneously we keep to our schedule of reducing the interior heat source by 3°C (5 °F) or so each week. It's a very gentle hardening-off process but we've never had a problem with it. We know that they're 'good to go' when the artificial heat source has been removed and the run is left open twenty-four hours a day.

At this stage of their development, on a hot day, we will occasionally 'mist' them with water for a minute or so. This is not a shower – just enough for them to feel the effects of the water. They start preening their feathers almost immediately, which is good for conditioning and gets them used to dealing with moisture.

As with the interior, we sprinkle the base of the runs with aviary sand, which it good for the chicks and helps with cleaning.

Feeders and drinkers are the same as in the nurseries, as are the toys, with the addition of the odd log placed here and there, which they love to clamber over and peck. CDs, either suspended from the cover or fixed to the side walls, and foil freezer containers provide further light entertainment.

During this first growth period we cannot emphasise strongly enough the importance of hygiene. Some minimum basic rules:

1. Clean out the nurseries/mini-runs at least once per week.
2. Clean food and water containers twice a day.
3. Water spillages and wet food must be cleaned straight away.
4. Poorly or dead chicks must be removed and treated, or disposed of in a sanitary manner.

We keep all food in heavy plastic bins with rodent-proof lids. Drinkers and feeders are religiously cleaned twice each day with fresh water. We use sterilising tablets where necessary, pot brushes and pointed-end paint brushes to access the tricky spots, and to get rid of the slimy residue. At the end of each nursery batch every piece of used equipment is disinfected using a specialist avian cleansing product.

Bits and De-beaking

As mentioned earlier, at around three weeks chicks may develop a tendency to feather-peck. I have already discussed prevention techniques that principally amount to making sure nurseries are not overstocked, and the use of props to keep them stimulated. However, if the chicks do develop the habit it is very difficult to stop. Feather-pecking causes injury and stress to the other birds, who then become susceptible to infection and disease through loss of condition.

We have been fortunate in having very few problems of this kind, probably because of the space we allow each bird and the fact that when we spot a bullying 'pecker' we immediately remove it (whether at nursery or rearing field stage).

Other than the preventive ploys previously described, there are three methods of dealing with the risks posed by feather-pecking. We use only the third as a precaution. The techniques can be used either as preventive measures or in response to an occurrence.

De-beaking. This involves using a special tool to slice through the upper mandible. The incision is cauterised at the same time to prevent bleeding. This permanently affects the form of the beak but does not compromise the birds' ability to feed or drink.

Bits. These are pieces of plastic or metal, usually in the shape of a ring. The bit fits between the upper and lower mandible with either end anchored in the bird's nostrils. The plastic bits are easier to fit and remove than the metal version. Using a 'bit-fitting' tool makes the job quicker, easier and less stressful for the bird.

Anti-peck compounds. There are several herbal and complementary products on the market specifically developed to suppress the feather-pecking instinct. These products usually comprise a mixture of minerals and vitamins, and are administered via water. You can also use a bespoke repellent to spray the tail feathers, but if you do, remember to wear gloves, it's a very penetrating odour!

If you decide to use one of the invasive techniques take great care and seek advice from your vet before you start. The whole process is unavoidably stressful for the birds so you need to be sure it is a necessary intervention. If you are in any doubt at all Defra, in its 'Code of Practice for the Welfare of Gamebirds Reared for Sporting Purposes', provides specialist advice.

The Rearing Pens

At around five weeks old the pheasants are transforming. The partridges go through exactly the same changes, but a little later, around seven weeks. They're becoming adolescents. Those of you with teenage children will immediately recognise the signs. If it's a boy, you're dealing with a home-grown thug. He loses his ability to speak and can only communicate through a series of grunts accompanied by a variety of shoulder shrugs depending on the message to be conveyed. The girls often become distracted and impatient, but more commonly sulky. The ability to 'answer back' has, by this stage, become something of an art form.

It's rather the same with game birds and features increasingly restless and quarrelsome behaviour, especially prevalent amongst the pheasants. Bouts of wing-flapping are followed by test flights up and down the run, which inevitably end in a crash-landing as they are not yet fully in charge of the controls. This results in frequent bouncing off the sides and knocking over anything in their path. These are the final signs that confirm it's time to move your birds out to the rearing pens, where they will live until they are ready to be transferred to the release pens.

Construction

Construction of the rearing pen needs attention to detail because it must be strong enough to resist predators, and idiot-adolescent proof. It is surprising how small a gap a young poult can get its head stuck in. With no idea how to reverse, it quickly tires and injures itself or dies. It can then, sadly, become an object of great interest to the others who are quite capable of reducing the corpse to a skeletal mess in a frenzy of pecks. Fortunately we have only experienced this once, but that was enough for us to redouble our efforts to make sure that the build was sufficiently fit for purpose.

The appropriate size of pens depends on your needs. But to give you an idea ours are approximately 15 × 12m (50 × 40ft) which will comfortably contain sixty pheasant poults, safely and without risk of overcrowding, up to their transfer to the release pens. As with the nurseries you should plan on giving them as much ground area as possible. This undoubtedly minimises the risk of bad habits developing, and also cuts down on the inevitable ingestion of parasites.

The construction of our pens resulted from lengthy research so we think it's

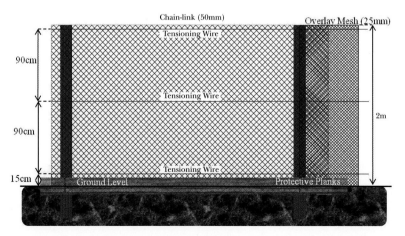

Rearing pen fence.

safe for you to use as a template. Modifications are, of course, fine, but the basic principles should remain if possible.

Materials needed are:

- Fence posts 2.4m (8ft) long.
- Bracing wire 3mm (⅛in) and tensioners.
- Planks 2.5cm (1in) thick, 20cm (8in) wide, 10cm (4in) of this buried into the ground.
- Galvanised main chain-link mesh with 50mm (2in) holes, enough for 2m (6ft 6in) height.
- Galvanised overlay wire mesh with 25mm (1in) holes, enough for 2m (6ft 6in) height.
- Small fencing pegs.
- Top wire mesh with 30mm (1¼in) holes or (preferably) game bird netting for the top cover.
- Planks to construct the door and frame.
- Assortment of suitable screws, nails, staples.
- Fence pliers for use with fence hog rings (essential).

N.B. *Always use the appropriate safety equipment when involved in a building project.*

The pen should be built in an area that does not have overhanging tree branches. Remember this is not a permanent home so if it ends up looking a bit stark it doesn't matter. Poults need lots of light and shelter from weather extremes, which can be provided by you. Overhanging branches just encourage winged predators, and risk damage to the netting cover. They can also contribute to 'damp patches' appearing in your pen – marvellous for parasites.

There are several schools of thought that say you should never build a pen that contains 90° angles. Instead it should be of 'sympathetic' design, being ideally oval or rounded at each corner. However, all our rearing and release pens are rectangular. We have had no problems at all with corners – but we do have with 'crannies', described later.

Starting at the baseline you need to dig a trench approximately 30cm (1ft) deep around the marked perimeter. The actual depth all depends on your base type. With our clay-based soil and its tendency to develop large crevice-like cracks during the summer and resemble claggy quicksand in the winter, we need the extra depth to make sure no gaps start to appear.

Enclosure construction stage 1.

Enclosure construction stage 2.

Sink fence posts every 3–5m (9ft 9in–16ft 6in) and staple bracing wire to each stake close to the bottom, middle and top, 15cm (6in), 90cm (3ft) and 90cm (3ft) respectively, and tension. Ideally line the perimeter at base level with planks sunk 10cm (4in) into the ground and reinforced with pegs. This allows extra protection against mammalian predators such as foxes and martens that can dig under the wire to access the poults. Fix the lower main mesh and lower overlay mesh and then fill in the trench. The smaller diameter overlay mesh provides double protection against predators but also prevents the poults from either hanging themselves or managing to fly through the larger diameter mesh. Then attach the upper meshes such that you now have a totally enclosed area up to the top of the fence posts, awaiting a roof.

Small diameter overlay mesh provides extra protection.

Enclosure construction stage 3.

You will now need to cover the construction. We used 30mm (1⅛in) wire netting, but you can buy bespoke 'game netting'. This is much more expensive but is made of a more 'forgiving' material that reduces the risk of accidental injury to cannoning birds, especially partridges that bounce around like shuttlecocks. In truth that would be your 'best practice' buy. The cover is strengthened by the straining wire, tensioners and supporting struts from within the pen itself.

Just for additional anti-predator security we roughly hammer a few 10cm (4in) nails part-way into the top of each stake. This doesn't look pretty but is a useful deterrent against winged predators like buzzards and sparrowhawks that, prior to this, we have often seen positioned on top of a stake looking for entry points. They don't do it now.

The gate is of conventional design using similar materials, and latch-locked. Do remember to bed in a threshold piece though, that again will guard against soil erosion and burrowing animals. We then surround the whole lot with a three-strand electric fence powered by a 12 volt lead acid battery.

Interior Furnishings

Using our rectangle shape we construct small hides in three corners, and A-frame shelters roughly in the middle, with the broad sides facing the prevailing weather. The hides need to be weatherproof so make sure whatever materials you use are suitable, and are not treated with toxic materials. In the corners we have used old doors as 'wings', and pallets for the front portion.

Interior furnishings.

We cover the pallets with tough plastic sheeting and overlay it with garden screening materials. We've experimented with bamboo, willow and reed but find the bamboo to be best because it is more resilient. It isn't expensive, and a roll will last you a couple of seasons. We then fix branches over the top of each just to minimise damage to the hide construction and also to give the poults a taste of Nature and something to play with. Branches are changed periodically because the leaves wilt and die and the whole lot gets soiled.

When assembling corner shelters you need be careful to fit the wings flush with the wire. If you leave small gaps or 'crannies' you risk a poult getting trapped between the wing and the boundary wire and injuring itself.

We also have a bathing area in each pen. Game birds love dust baths. It is a method of getting rid of the mites that live on their bodies so it is important to encourage it, especially after a spell of bad weather. Again, this will depend on your ground conditions but we clear an area of bare earth at the highest point which we cover with a mixture of sand and gravel.

We also erect several perches to get the birds used to the idea of roosting; a couple of bales of straw set on pallets (to keep the damp off them); and several logs. If it rains we cover the straw loosely with a sheet of thick black polythene pinned down by logs. This keeps the worst of the weather off and stops the straw going mouldy. The rest of the ground area is grassed or multi-seeded.

We try to follow a 'fallow farming' system and always allow at least one pen a year's rest. After the birds have been transferred to the release pens we will rotovate the ground and then broadcast seed mix. In the following season we will cut it right back just before the birds are brought in. Pheasants flatten tall grass and crops very quickly and partridges get stuck in it, so it serves no useful purpose to maintain a jungle. The freshly cut plants, on the other hand, seem to work very well in 'holding' the soil, and give the birds a good natural base. Although this is only our theory, we think the plant roots may also minimise parasite activity. If this is, in fact, correct, it is perhaps because the birds cannot access the bare earth so readily.

When the poults first go into these pens we take their original feeders and drinkers with them. These are the small trough mangers with anti-scratch bars, and galvanised iron or fount plastic drinkers. They are used to these, so know what to do. Since they are not famed for being overly brilliant it is counterproductive to cause widespread confusion at this stage of their young lives by introducing alien-shaped containers.

However, we make the transition to auto-feeders and water hoppers as soon as possible by running the two systems in tandem for a few days until they've understood the changes. The rearing pen feeder we use is the same 'netball goal' design as used for the parent birds. The advantages are that the food in the container is kept dry at all times and, despite the fact that there is a tendency for them to peck at the basket, just to play, there is, on the whole, less wastage. We place a lump of wood or old slab stone beneath the basket to avoid the birds pecking directly on the earth. It's also much easier to keep the area below the basket dry, clean and tidy. In addition we regularly mix in bespoke bird grit, which is usually a combination of finely crushed oyster shells, sand and often an attractant like anise. The birds can't help ingesting a little whilst they are feeding. All this is designed to help their digestive systems function properly, stimulate healthy bone growth and encourage them to eat the right rather than the wrong things.

General Routine

Our routine, until the poults are old enough to be transferred to the release pens, is just the same as with the parent birds:

1. Feed and check water twice per day.
2. Provision of preventive medicines and vitamins in food and/or water as required.

3. Prompt removal and treatment of sick or injured birds. These are put in the 'hospital pen' (see Chapter 6).

4. Receptacles cleaned thoroughly. There is no need to develop an obsessive compulsion about this but the containers must be cleaned regularly – especially the drinkers.

5. Rearing pen maintenance, which includes the following checklist questions:
    - Is the electric fence battery working?
        - Do we need to put down weed-killer?
        - Do we need to remove any branches or debris from the cables?
    - Are there any changes that need to be made inside the pens, such as a fresh set of branches or removal of deep litter materials after a bout of bad weather?
    - Are there any holes in the fencing materials?

The birds will remain in the rearing pen until they have developed sufficiently to be transferred to the release pen. This will be their last stage of close care before the start of the season.

4

RELEASE PENS

'THE GREAT OUTDOORS'

Our timing to move the poults from the rearing to the release pens depends on several factors and differs between the two species. I have separated the techniques we follow by describing that for pheasants first, and then the process for partridges. Each system does contrast for important reasons, which I explain as we go through.

Release Pen Principles

The purpose of a release pen is to temporarily 'hold' the poults on a site that is convenient for your shoot. This is a very short period for partridges as I describe later. Pheasants need longer to be able to acclimatise in the relative safety of their enclosure, enabling them to gently explore the surrounding cover crops and topography as they develop. It becomes their home. The pens will vary in size and number depending on flock size and land availability. Just don't forget that the more you build the more you will need to maintain. We built a small number of large pens which reduced the overall workload. The pens will act as a focal point for the drives when shooting starts. As David

Hudson states in his book *Gamekeeping* (Swan Hill Press 2006): 'In theory – though not always in practice – pheasants will tend to fly back towards the release pen when they are flushed, thus allowing a line of guns to be sited under the flight line.'

The 'Gentle Release' System

We have two types of release pens. This sounds unnecessarily complicated but it results from our evolution of the process. We rarely use this system now for pheasants, but will do so for partridges in a modified form in the 'woods' pen described below.

When we first began rearing game birds we kept both the pheasants and partridges in the same pens until final release. We were only rearing small numbers so until we were sure we wanted to continue with the shoot, it didn't seem worth the extra expense to build separate release units. Instead we followed the 'gentle release' system.

The idea is that you keep your poults in the rearing pen until they are old enough to be released, and adapt the same pen for that purpose. There are two ways of doing this:

1. Using the rearing pen construction as described in Chapter 3, you can lift and peel back the netting 'lid' to allow the birds to fly out and return to roost.
2. Alternatively, the pen gate is opened so the birds can come and go as they please.

In both cases food and water is continued in the pen for a couple of weeks after the conversion whilst the birds are exploring outside. The theory is that they will use the pen as home until they naturalise on the land.

One of the key considerations of either system is the pen location. In order to be successful as a final release mechanism they need to be built on sites that are suitable for your shoot.

We originally tried both systems (using three pens of similar construction), with differing results. Two were built within an old disused orchard not far from the farm buildings. A couple of trails led from the orchard boundary to the main woodland cover areas and these became suitable rides. The whole site was surrounded by a rickety post and rail fence, which we patched up and reinforced with electric fencing. The third pen was built roughly in the centre of our land on the edge of the woods, adjacent to pastureland. It was the zone designated for our rough shoot.

For the 'orchard' pens we just propped open the doors. Feeders and drinkers were left inside for around two weeks, and then placed directly outside. We positioned others throughout the orchard. Even though they were not ideally situated for shooting purposes, the system itself worked well. Most of the birds were perfectly happy to remain in their immediate surroundings for the first few weeks before exploring further afield. They were safe from the risk of immediate fox predation because of the protection afforded by the electric fence, and the cover on the pen sheltered them from avian attack. It also worked because the poults could potter about on foot and eventually find their way back in through the open door. In fact we are sure that the group of pheasants that live around the fringes of the buildings today came from these early batches.

Common buzzard. *(Photograph courtesy Laurie Campbell)*

We occasionally still use an adapted version of this 'gentle release' system for pheasants to 'mop up' birds that are simply too young or underdeveloped to go to wood with the older poults. This can occur for a variety of reasons, including the parent birds themselves producing a clutch late in the season, or the broodies getting to work a little late, or even the final incubator batch. Whatever the reason, if we feel they are just too immature to go into the 'open top', then we will use the other method as a fall-back.

The second method we initially tried for pheasants only in the 'woods' pen. It did not work. As instructed, we peeled back the wire netting lid but kept the door closed. We positioned ourselves in a pop-up hunting hide about 80m (88 yards) away to see what would happen. Eventually one or two poults clambered out, and then about twenty more. Within the first thirty minutes a buzzard flew into the pen and attacked one of the hens, killing her very rapidly. Bearing in mind that these raptors have eyesight eight times keener than humans, we should not have been surprised at their optical vigilance. It was incredibly frustrating to observe this legally protected species at work without being able shoot it. Most of the poults eventually flew out and explored the adjacent ride. Two more buzzard attacks came that afternoon.

The other problem was that, once out, most of the birds could not work out how to get back in again. They continually circled the pen on foot, looking for an opening. The experienced amongst you will by now have eyes cast skyward thinking: 'Of course that's going to happen, you idiots!' And you'd be right. We concluded that it was foolhardy of us to expect these young birds to understand how to fly back into the pen having haphazardly scrambled out.

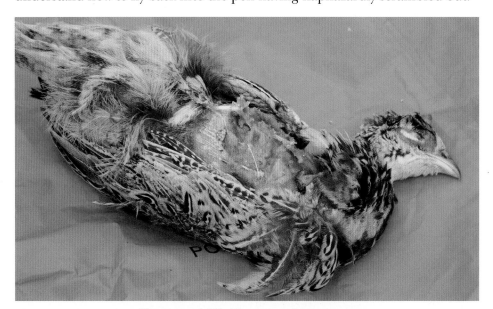

Pheasant poult killed by common buzzard.

Working with the birds as they developed should have already taught us that. They do almost everything on foot and only take to the wing when really necessary. As they get older the problem is likely to compound itself since, being relatively heavy, a quick sprint using a good, strong pair of legs is probably less energy-consuming than flying. Clearly doomed, this experiment was cut very short. Unfortunately, having made several failed attempts at re-entry, quite a few poults wandered off in dribs and drabs. Inevitably our attempts at herding them back became very difficult, and created a further opportunity for predators.

So, we are reasonably happy with the 'opening the door' method (with the addition of 'pop-holes', described below); we do not recommend 'lifting the lid' as a release system.

The Open-top Pheasant Release Pen

The big differences with this type of release system are:

1. Bespoke separate pen with no roof.
2. The pen is intended to incorporate the natural habitat – with some modifications. It's really a mix of natural and some artificial cover.
3. The birds should regard it as a 'safe haven' and roosting site rather than a temporary home.
4. At ground level it has fox-proof re-entry points (pop-holes) positioned at regular intervals. These are intended to direct birds back into the pen that have flown out and can't work out how to fly back in again.
5. It has an electric perimeter fence surround.
6. It's much bigger.
7. Ours also has a small, fully enclosed, pen in the centre.

The construction method and materials needed are very similar to the covered rearing pens discussed in Chapter 3. The difference is that you do not need to use the lower protective planks, but you will need to buy the extra materials to build the pop-holes.

To help give you an idea of dimensions and poult capacities here is an extract reproduced with kind permission of the Game and Wildlife Conservation Trust (GWCT) from their 'Code of Good Shooting Practice'. It outlines best practice in each area of the sport and offers practical advice. It is regularly reviewed by a steering group comprising the most notable organisations in the industry. Albeit aimed more towards the flora than the fauna side of things, it nevertheless provides a useful guide:

Releasing Pheasants and Partridges

- Shoots should refer to the Game and Wildlife Conservation Trust guidelines for sustainable game bird releasing.
- These provide a rule of thumb, advising that in order to avoid damage to habitat in most situations shoots should avoid releasing more than 1,000 pheasants per hectare of pen, and more than 700 per hectare of pen in ancient, semi-natural woodland and that pheasant release pens should not, in total, take up more than about one-third of the woodland area on the shoot.
- Where shoots exceed the recommended densities they should be able to demonstrate that their particular circumstances and management regime (for example, by limiting the period of time birds are in release pens) does not significantly damage woodland flora and fauna.
- Partridge release pens should be sited in cover crops on arable or improved grassland, rather than on semi-natural or unimproved habitats.

The Site

Think carefully about the site for your release pen. To minimise losses to neighbours, common sense dictates that you need to build it as close to the centre of your land as possible, whilst being convenient for your drives. That said, the site ideally needs to have a south-facing aspect and contain a mix of suitable woodland and some pasture. Evergreens, e.g. pines, are an ideal tree type, allowing year-round roosting capability and good protective cover. Pasture and bare areas are also essential for the birds to dust-bathe, feed and dry off after a bout of wet weather. Mix these with areas of ground cover to give the birds protection from stormy weather and to provide places in which to hide from winged predators (and from one another when an argument breaks out).

Try also to avoid building where there is a natural source of running water. This is fine adjacent to cover crops but, for good reason, not inside the pens. There is always the outside risk that a brook/river becomes an entry point for small predators, and whilst it is undoubtedly extremely tedious having to check and replace water very regularly, it is the only reliable method of administering medication, vitamins and minerals. As I explained earlier there are several feeds available on the market containing medication, but in this setting the food intake of each bird is less reliable to predict and monitor, especially in a pen with natural food sources. On the other hand, they do have to drink.

In constructing our release pen, whilst we were able to select a site more or less in the middle of the estate, it wasn't perfect. It featured a predominance of pine trees which were good, but they made the whole area heavily shaded. This meant there was very little ground, or intermediate, cover. To solve the problem we felled several poor specimen trees and simply left them in place. That immediately gave us what we needed, more light and intermediate cover. We know the fallen boughs won't give cover forever, but they will do the job adequately for a few seasons. Once sections begin to 'compost down', they provide a haven for all sorts of insects and grubs. These, in turn, are a natural food source for the poults and any chicks that are produced in the following year.

Construction

Much has been written about the best designs for release pens but the one we favour was described by Jamie Stewart, who published his recommendations in the BASC's magazine, *Shooting and Conservation*. Although ours ended up being a little different, his advice, reproduced below with the BASC's kind permission, was practical and easy to follow. The perimeter, which in our case encloses 1 hectare (2½ acres), was constructed broadly in line with Jamie's recommendations:

> A track should be marked out and cleared prior to construction. It should be wide enough to include pop-holes with 'wings' and allow the keeper to walk both inside and outside the perimeter fence for daily inspection once the poults arrive. All branches overhanging the track should also be removed to discourage poults from flying out of the pen before they have become acclimatised and to prevent easy access for mammalian predators.

Surplus wire bent outwards to stop small predators scaling up and over the fence.

The main chain-link perimeter fence needs to be layered for added protection. A single 1m depth covering of 30mm (1¼in) diameter netting fixed at ground level is recommended. We doubled the height and used two for extra protection. Starting at base level we fixed one above the other, which gave us a height surplus. We bent the remaining material outwards to stop small predators scaling up and over the fence.

We agreed with Jamie's recommendations to run a two-strand electric fence around the perimeter. Our base cable is set 15cm (6in) off the ground and the second is 15cm (6in) above this. Instead of erecting a separate fence as suggested, we attached the cables to the boundary fence stakes using 10cm (4in) insulators. This allows the cables to protrude a sufficient distance from the main enclosure without shorting on either the main fencing posts or the chain-link itself. It was a cheaper, simpler option, and works well with our pop-hole design. Once built, we treated the area with a strong weed-killer to prevent foliage shorting the conducting wire.

Our pop-holes/anti-fox grids are built to last and again, are based on Jamie's original design ideas. To use the same design you will need the following materials:

- Round iron bars 6mm (¼in) diameter or 8mm (⅜in) if you're feeling strong
- Welding equipment
- Angle-iron 25 × 25 × 3mm (1 × 1 × ⅛in)
- Overlay mesh 25mm (1in)
- Tent pegs

They feature a fairly conventional metal grid that fits flush to the main fence. We made these ourselves at a fraction of the cost of buying them in. They measure 30cm (1ft) high with 9cm (approx. 3 ⅜in) gaps between each bar and 6cm (2 ⅜in) base spikes that bed into the soil. Outside the pen we have 'wings' formed as an arrowhead shape pointing towards the grid to help guide poults back in, and let's be clear, some of them need guidance. The electric fence cables surround each pop-hole. They are fixed to small stakes at the splayed section of the 'arrow head', and feed back to the perimeter insulators either side.

Pop-hole material purchases should be based on fitting one at a maximum

Pop-hole frame made from iron bars and angle-iron.

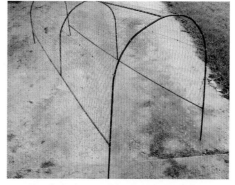

Pop-hole frame with wire mesh overlay.

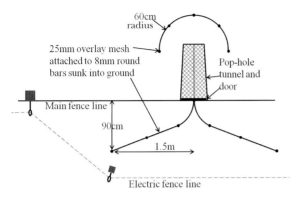
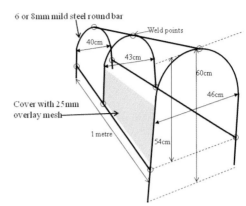

Pop-hole construction.

Pop-hole tunnel.

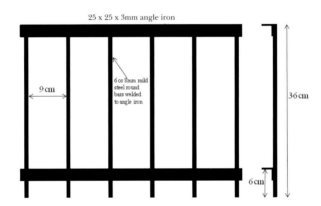

Pop-hole door.

distance of every 50m (55 yards). We say this having already observed many birds endlessly patrolling the same short stretch trying to find a way back into the pen. If they are unable to do so they are at risk of predator attack against the fence, or they may wander off aimlessly into the woods. So, if you want to retain as many of your game birds as possible, it simply makes sense.

Fortunately, pheasants are almost completely unaffected by the current from electric fencing so this does not impede their entry in any way. They just step over it. On the inside of the pop-hole we have an entry tunnel linked to the perimeter fence and grid. The tunnel was formed using three mild steel metal hoops that descend in size, the largest being directly adjacent to the grid. It resembles a lobster pot. Each was welded together using a single spine of the same material. We then covered them with 25cm diameter netting. To prevent

Pop-hole fitted to release pen.

the poults from using these as doorway exit points we followed convention and fitted a 1.2m (4ft) diameter 'U' shaped piece of netting fixed by iron pegs around the internal opening of the tunnel.

When you are making the entrance to the pen make sure the threshold is dug in and think about the gateway dimensions. Ours are wide enough to drive a quad bike/small poult van through.

Having built the main perimeter fence we needed to sort out the interior. For the main feed ride we followed accepted practice by cutting a serpentine shape in the central part of the pen. It is 'S'-shaped because straight lines full of poults feeding make much easier targets for winged predators. We positioned other feeders and drinkers towards the edges (see below).

We decided to build a small multi-purpose aviary in the centre of the pen. This is completely enclosed and it currently houses a couple of broody hens and their pheasant chicks. We also have four cockerels who hang around outside.

Guardian cockerels.

Multi-purpose aviary.

The idea here is to use these birds to attract the others back. We release the poults when they are old enough to survive in the release pen, and keep the chickens enclosed, with a cockerel back in residence. This causes the poults to stay near the aviary just to be close to their surrogate mums. The general poultry activity in the aviary is also fascinating to many of the other poults, who frequently examine the inhabitants with great interest. We believe that this contributes to helping 'hold' the poults for longer. Following the same objective, future plans for the pen include keeping a small harem of pheasants to act as 'call birds'.

The cockerels are there because they are extremely effective at signalling potential danger. The poults, who have been brought up alongside them, seem in part to respond positively to this (although sadly not frequently enough), and will often scurry for cover when they give their warning call. We have never had any bullying incidents from the cockerels, but we did take care only to use docile birds.

Introducing the Birds

Once the release pen is built you are ready to transfer the birds. Regardless of whether it is a new pen or not, it is good practice to do a last-minute check by walking the whole site. Be on the lookout for all sorts, ranging from stray bits of debris left from the building process, to holes, or even unwanted visitors who may have set up camp in the undergrowth. Make sure your rides are the correct shape, not dead straight, and cut back nicely. And last but not least, make sure the food and water supplies are fresh and located evenly throughout the enclosure. Use the same type of feeder and drinker that the birds have been used to in the rearing pens, and make any changes gradually after they have settled.

Poults will instinctively gravitate towards the sunny spots, and then the perimeter fence. You need to make sure you respond accordingly by positioning food and water in these areas. We built several traditional slope roof shelters which were situated throughout the pen bordering the main rides. We matched them up with a feeder and drinker. To keep costs down you can easily make these using appropriate materials but don't forget continuity for the poults and their ability to get flummoxed. That's why we plumped for the traditional kit they knew.

Sloped roof shelter.

Standard tripod drinker.

Your next decision will be the age at which to release your poults. If you have opted to buy-in, the decision may have been made for you, but there are relevant arguments that could persuade you to purchase younger or older birds if you have the choice. Of course you won't be surprised to learn there is no exact optimum for this. Equally, by now, you would be a little disappointed if you felt every expert shared the same opinion. Never fear, strongly disparate views are certainly held. Listed below are some of the general pros and cons, followed by what we do.

Releasing pheasants at six or eight-plus weeks

Cons for releasing at six weeks:
- Still relatively underdeveloped and often vulnerable.
- May not be fully feathered.
- Poor flyers.
- Not at all 'street wise'.

- More susceptible to a greater range of predators.
- Arguably less well equipped to withstand extreme weather.
- Often not yet fully adjusted to eating grower pellets and wheat.
- Less robust health-wise.
- Not adept at roosting in trees.
- Therefore a potential waste of money because it's just too early.

Pros for releasing at six weeks:
- Often adapt better to new environment.
- Less likely to stray.
- Relieves space in the rearing pen for a new batch.
- A popular choice with game farmers.

Cons for releasing at eight-plus weeks:
- More likely to wander.
- Higher risk of disease outbreak because they have been kept closer together for longer.
- Less likely to 'home' well in new environment.

Pros for releasing at eight-plus weeks:
- Stronger, bigger birds.
- Fully feathered.
- Better resistance to bad weather.
- More likely to roost in trees.
- Less susceptible to smaller predators.

In fact there is no 'perfect' release age so it becomes personal preference because there are some very logical reasons for and against each.

We err towards the older poult, and generally transfer at eight-plus weeks. Our key determining factors are the general health of the batch (we keep any slow developers back), and the weather. There is absolutely no point transferring a group of poults to the release pen in foul weather. They take time to acclimatise, and are at risk of going downhill fast and becoming highly stressed if conditions are poor. A really bad night of cold and wet can easily kill a high proportion of your poults if they are exposed to it immediately after release, so do your best with forecasting and act accordingly. This is also an extremely important issue if you buy-in your poults. Whatever the age at release, make sure the birds are fully feathered and hardened off, and work with the game farm to try to receive them during fine weather. If this means collecting them yourself, or delaying delivery, then do so. Other genuinely important factors for us are rearing pen space needs, and our time availability for release pen keepering activities. But these have to be secondary concerns.

Wing-clipping

Wing-clipping is a further consideration for pheasant poults (but not for partridges for reasons given later). It may be done at six weeks and involves removing the juvenile primary feathers (the first seven or nine) on one wing. The theoretical advantages are that by temporarily forcibly inhibiting the bird's ability to fly it remains in the release pen longer than it would otherwise do instinctively. Wing-clipped poults are less likely to fall as early prey to predators like foxes, and have time to become accustomed to their new regime of feeding. They therefore 'home' more successfully than many others. Some gamekeepers will mix and match by using the technique solely for birds that are put in release pens situated close to a boundary. It is a painless procedure that usually lasts between four and six weeks depending on the speed of the re-growth.

The counter-argument is equally logical. Why do something to extend the birds' dependency on humans at exactly the time when we should be distancing ourselves from the flock, and preparing them for survival in their natural habitat? This is a real concern for many keepers, who are worried that the knock-on effect of this practice may impair the birds' ability to give a good show for the shoot. These birds are not pets. Wing-clipping also deprives a bird, albeit temporarily, of natural behaviour and some protection from the elements. A final point to bear in mind is security. If your pop-holes are open there is the distinct possibility that a wing-clipped poult may find its way out, and it is then at a real disadvantage in the event of a predator attack.

When we first built our pen we decided to try both systems. We released four batches of around seventy poults into the same pen. We wing-clipped the first group and closed off the pop-holes. They stayed well and (although still exposed to risk from avian attack), fluttered about happily inside the pen protected from mammal predation. We did not clip the second lot. This meant we needed to re-open the pop-holes but that did not seem to be a problem at all for the existing birds. Those that did manage to find their way out just came right back in again and anyway their flight feathers were growing back rapidly. To our knowledge we have never lost any poults to predation as a result of being temporarily unable to fly properly.

Interestingly (read frustratingly), we had significantly more early problems with the second group who, despite their enormous pen, decided to explore outside very quickly. This resulted in several predator attacks, often during the night, in the vicinity of the pen with poults having flown out and not managed to get back successfully. Our decision was made. The following two batches were duly clipped.

A Note on 'Batches'

I keep mentioning releasing 'batches' of birds in the same pen. As you will recall, this situation arose because of our relatively small incubator capacities and brooders. For sensible reasons the very notion of batch releasing in the same pen will cause many experienced keepers to recoil in horror. If you bring new batches of birds into an existing flock you risk introducing disease and causing disruption to the current incumbents. Medication programmes (see below) are thrown out of kilter, and opportunities for outbreaks of bullying may emerge as the older ones pick on the youngsters. However, we have not experienced any problems at all with this form of release. But there are some specific reasons why. Our birds come from the same closed breeding flock, and are raised in exactly the same way, so we do not have the problems often associated with introducing 'externals'. Therefore the introduction of disease is far less likely.

Medication

In the rearing pens all the birds are routinely wormed as part of our preventive treatment programme. During the batch transition period (usually over a four- to five-week period), the release pen birds are given a monthly dose of 'Verm-X' herbal mix in their water. Active ingredients include garlic, peppermint, common thyme, cinnamon, quassia, tansy, cayenne and nettle, none of which conflicts with any chemical treatment that may be necessary. One week after the last batch is released the whole group is dosed with a veterinary prescription anti-gapes mix, and a coccidiostat treatment. As for bullying, fortunately we have never had problems and that is probably because the release pen is designed to hold at least 700 birds, and we have around half that number so – lots more hiding places.

'Holding' game

We also believe our method can have distinct advantages. Consider the other end of the process. The idea is to be able to rear enough game birds to sustain a worthwhile shoot. It's not always that easy. Since one of the real challenges is being able to 'hold' the game long enough for it to naturalise on your land, or at the very least remain close to the release pen for as long as possible, the periodic introduction of new batches of poults certainly does act as a natural attractant. We have observed on many occasions older birds coming back to the pen to investigate the newcomers. This, incidentally, is how we acquired our melanistics. Of course we do have mixed ages when it comes to the start of the shooting season, but that's one of the reasons why we harden off our birds as quickly as possible so that even the youngest are 'good to go'.

Survival skills?

The other question that occurred to us was whether there is any evidence to suggest that pheasants learn survival skills from one another. We assessed the effects of predation within the release pen. Out of the four batches we released over a period of about five weeks, we found that the third batch almost immediately started running between cover points. Where they copying their mentors? Conversely, the first batch lounged around enjoying the sunny but exposed areas and seemed to forget, in the general euphoria of things, to look skyward. Ironic really, since this group homed beautifully. Consequently, despite ample warnings from various cockerels, we did suffer some losses to raptors attacking inside the pen. The second batch was similar. Both groups learnt the hard way and eventually began to respond more quickly to the cockerels and run for cover when the need arose. With this behaviour now established, the newcomer third and fourth groups seemed to just follow their siblings. They probably didn't know why, but we genuinely feel we had fewer avian predator losses from these last two batches.

Of course without tagging each bird it's hard to determine precisely which ones are lasting the longest, but it's not that difficult to *estimate*, given the development and feathering of the birds. Our cautious conclusions are that, in the release pen setting, the youngsters appear to learn by rote, which may make the 'safe' batch release system worth using as a specific technique, or at least giving it further consideration.

The Transfer

Regardless of whether you are rearing or buying-in, this process begins from mid-July. Transfers should take place very early in the morning. This gives the birds the whole day to acclimatise to their new surroundings. If we are wing-clipping we do this as we catch each bird. This means they are only handled once and the wing is 'freshly' shortened. The catching is done by my husband and I, the benefits being that there are no excuses for extreme histrionics from the birds because they already know us well. Equally, by now we understand their flock foibles, some groups being prone to more flighty behaviour than others, so we can adapt accordingly. Our rearing pens are relatively small, so we either catch them by hand in the corners or using a net.

If netting, we use a micro-mesh telescopic 'catch and release' landing net. It has a spongy rubber-covered 50cm (20in) diameter head frame, fitted with fine soft nylon netting and a 1.6m (5ft 3in) telescopic handle. The materials are all soft but durable. Don't forget that, at this stage of their development, the poults are surprisingly strong and will easily snag nets with their beaks or claws if the material is too flimsy.

When you are catching poults, do not hurtle around the pen flailing your net above your head in 'fly-casting' mode. You will either panic the birds, or snag and rip your net on some of the interior furnishings. Single out one group at a time, move towards them gently with your net at hip height and pop it over a likely candidate. Swift and efficient action is called for at this point, with your partner ready to collect the bird, keeping its wings close to its body. Any dithering from either team member risks the poult injuring itself by accident, or escaping back to the flock. If the bird does panic during this process, place your hand over its head and lightly cover its eyes: it will immediately become subdued. Once the bird is extracted from the net, the wing is gently extended and offered up to be clipped. We use a sturdy pair of kitchen scissors to do this job. We then place each bird into a bespoke game bird carrier, stick it in the shade, and continue until the job is done.

The birds are then transferred to the release pen, which in our case is very close. We place the carriers more or less in the

Wing-clipping a poult. *(Photograph courtesy Solway Feeders)*

centre of the pen in a large circle. They are close to the main feed ride, and to secure ground level cover points. We raise each lid smoothly and let the birds come out by themselves. It's important not to frighten them at this stage. Normally the first few are a little timid, but they soon pick up and start to bustle about searching for food.

Husbandry Post-transfer

In terms of husbandry, your routine will be very typical from here on until the beginning of the season. It involves a visit to the release pen at least twice each day. Food, water, medication and nutritional supplements must be supplied regularly. During this period you will need to change the diet by introducing wheat and, a little later, maize. We start this off early and give sprinklings of each to the poults during the latter stages of their development in the rearing pens. The idea is to gradually introduce them to more natural foodstuffs so that when they are fully released their eyes and guts are accustomed to looking for and digesting the real thing.

At this stage of their growth the birds are young and inexperienced, and will probably be more excited by their new surroundings than by the food and water. They will recognise you so, despite the fact that they must not be treated as pets, it is helpful to use a call of some kind to attract them to the feeders. Many keepers will whistle the birds in whilst others use a voice call. Ours are attracted by the sound of the quad bikes coupled with a gusty bellowing of: 'Are you hungry?' Whatever works for you and them is fine, but if you decide to go for a handheld gadget, don't forget it. 'Calling back' is also a handy device for bringing them back in before a shoot, and helps the 'dogging in' process described later.

Part of your routine will include a regular patrol around the interior. Look for signs of disease and death. If there are any unexplained deaths consider taking the corpse to your vet for diagnosis by autopsy. Of equal importance is the routine maintenance to the release pen itself. You must check both the interior and exterior perimeter fences. Do the electrics still work? Have any overhanging branches appeared that are perfect launch pads for raptors? Is all of the fencing robust and hole-free?

Quad bike 'pack horse'.

The Partridge Release Process

Partridge juveniles do not adapt well to 'open top' pens. If you decide not to follow the gentle release method described earlier, you need to adopt a completely different, more standard process. Partridge pens are much smaller than those for pheasants, intended for temporary use only and usually made in completely knocked down form. A typical example is shown in the accompanying photograph kindly provided by Solway Feeders. The pen measures 6.1 × 3m (20 × 10ft). It was used to successfully release thirty red-legs during the 2011 season, with a sheet of corrugated iron and branches inside for shelter and cover, and for a smaller number of greys in the 2012 season.

Partridge release pen. *(Photograph courtesy Solway Feeders)*

In many ways the same release guidelines apply for both the grey and red-legs but there are some important considerations. I sought advice from Dr Mike Swan, Head of Education for the GWCT, on the process and he made these comments:

> If you want to release some greys just for shooting, you can do much the same as with red-legs, but returns are usually very low. Also, since they are much more inclined to stray than red-legs, you can swamp out local wild stocks with what may amount to poor-quality birds. This is hardly fair on any neighbours who might be trying to help restore a small wild stock.

So the question of effective release can be far more complex, especially if you have neighbours that fall into this category. If, as many gamekeepers do, you have a dual ambition to provide a successful show for the Guns and encourage the regeneration of wild stock at the same time, unless you already possess the requisite knowledge it is undoubtedly better to seek advice from an organisation like the GWCT, a recognised authority in this area.

Because partridges are birds that form coveys they ideally need to be kept in suitably sized groups that are more natural for them. This will usually mean making several pens and never only one for each drive. As the Solway Feeders example shows, you can keep several poults together in one release pen, up to a recommended maximum of fifty. It all depends on the number of birds you intend to release, and the number of drives you have on your shoot. The combination of these two factors will determine the size and number of pens you make. Bearing in mind that this is a relatively quick release process a pen doesn't need to be huge, just predator-proof and portable.

There are two advantages in having temporary release pens. First, you can put the birds exactly where you want them across your intended shooting area, which is helpful if you want to change the site each year. Second, by having a new patch of ground each year there is limited risk of parasite build-up, which is almost inevitable in the permanent release pens. If the release process is followed properly, there are no additional benefits derived from holding partridge poults for a long period.

Partridge Release Pen Construction

When building partridge release pens there appears to be no definitive advice on space allowances for each bird. But bearing in mind that they are only used for a few days, working off a minimum calculation of 30sq cm (1sq ft) per bird is perfectly acceptable. If in doubt, our advice would be to err on the side of a larger structure.

Designs include a basic square, or 'A'-frame collapsible structure with the following common features:

- Lightweight portable boarding approximately 35cm (14in) in height around the baseline.
- Sides covered with wire netting approximately 25mm (1in) in diameter.
- Soft top net to prevent the poults flying out (bespoke avian netting is best and avoids inadvertent suicides).
- Under-netting (double-layered chicken wire), if the poults are to be kept

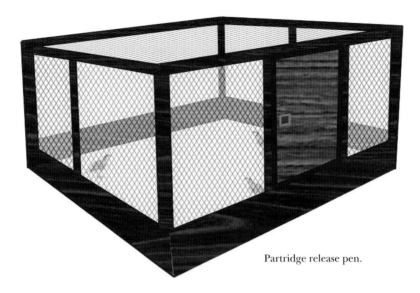

Partridge release pen.

longer than twenty-four hours. This prevents predators from digging underneath. Otherwise leave open for them to peck freely at the grass, but peg the structure down.

- Fit a couple of handles so it can be carried easily.
- If the pen is small make a simple hinged door that is easy to operate, causing the least disruption to the birds.
- Electric fence around the perimeter.

The interior of the pen should include a drinker and feeder and some kind of protection from the weather. Evergreen branch shelters are ideal. The addition of something to use as a vantage point is especially attractive to red-leg partridges, who will use it to call to those already released, or the other penned birds. A bale of straw or stack of pallets works well.

Release pens should be sited within a suitable game crop. The GWCT states: 'Red-legged partridge release pens should, wherever possible, be placed in game cover planted on arable or improved grassland, rather than on semi-natural or unimproved grassland sites.' They do have a tendency to gravitate towards woodland at the back end of the season but for obvious reasons these will not be suitable for the pre-season release site.

As an interesting point on conservation, the GWCT also notes: 'Avoid placing release pens or partridge feeders next to high conservation value hedgerows. Allow a buffer zone of game crop or other cover to keep concentrations of birds away from the hedge.'

Once you have selected your site you need to prepare feed, shelters and

Feeder protected by rough post and rail corral.

drinkers on rides areas immediately surrounding the pens to help hold the birds. We position feeders and drinkers at the base of dense evergreens. We have several deer in the area so to protect the kit we build rough corrals made from 1m (3ft 3in) stakes with planks attached. This looks like a miniature post and rail fence, and allows the poults (of both species) in, whilst keeping the larger game out.

Introducing the Birds

Partridges mature more slowly than pheasants so you should not be thinking about moving them from the rearing field to release pens until they are ten-plus weeks old. This means that you will be releasing the birds in August. To catch them up in the rearing pen we have two methods. Remembering that they are generally more nervous than their larger cousins it is important to use a technique that causes the least disruption. Since we only release small numbers, the most effective system for us is using a double-door box trap. Again starting early, we herd them along the fencing and into the traps. They trip the central treadle which drops the doors either side. We quickly cover the trap with a dark sheet and feed them into the bespoke bird transport boxes. The alternative is to use the same landing net as with the pheasants but, given the small numbers involved, we prefer the traps.

Just on a point of clarification, never wing-clip partridge poults. This form of release does not lend itself to having a bird that cannot function naturally. Your birds need to be able to hit the ground flying.

Partridge release takes place between two and four weeks before the start of the shoot and always during a period of mild weather. If you have a mixture of grey and red-leg partridges, keep them in separate release pens. When releasing partridges you need to do so in batches. We know they are social birds and they will retain the instinct to remain as a covey. Keep them all together for between two and ten days, and then start letting a few out at a time. (It's hard to be precise about time periods because each keeper has their own preferences, but there is certainly no evidence that suggests there is anything to be gained by keeping the birds penned for long periods. We begin releasing on the third day, the theory being that they have had a couple of days to settle down as a group, but not too long to feel the stressful effects of relatively close confinement.) Since they are prone to panic, you need to conduct this process very gently. Some keepers have a rope attached to a latch on a trapdoor which, when lifted, allows them greater distance between the birds and freedom. Of course the risk with this practice is that you have limited control over how many actually come out in one go, and may have to contend with a mad dash brought on by a crow sneezing in the next field.

You will probably witness them disappear into the sunset, but don't be alarmed. They do usually return fairly quickly to the central group, attracted by a dominant cock bird calling them back. The idea is to 'drip feed' this process over a few days in the run up to the shoot opening. If possible, keep a couple of 'call birds' back and release them just before the season opens. Some keepers will release all the birds in this way and others will try to keep a pair in the pen to serve as an attractant throughout the season. We would not recommend this if the pens are situated directly in the drive areas. It is too stressful for the birds left literally in the firing line and for obvious reasons goes against all the advice on the subject under various codes of practice. Ideally all your birds should be fully released just in advance of start of the season, which allows you to pack up your kit, clean it down and put it into storage for next year.

'Dogging-in'

As the poults develop, so will their natural instincts to explore. Cover crops, discussed in Chapter 7, are enormously helpful in retaining your stock, but the birds still need some encouragement and 'pointing in the right direction' if their wandering instincts get the better of them. Some, often the hens, will naturalise quickly and if you have a large open-top pheasant release pen you may well find some nesting inside the following season. The partridges, of course are 'quick-released' but they will generally have a greater tendency to remain in the locality so long as their needs are catered for.

Unfortunately there is no doubt that cock pheasants in particular will roam, probably because they are searching for a territory of their own. However, if this is allowed to go unchecked they will inevitably take a string of hens along with them and before you know it you're providing your neighbour with a great couple of days' free shooting and Sunday lunch.

Feeders and drinkers strategically positioned around the mid-point areas of your land undoubtedly help combat the wandering instinct. If you see your birds roaming beyond to the outer limits of your territory, try to spend a little time each day in the early weeks after release to call and chase them back towards the release pen – which they should by now consider 'home'.

Many keepers spend a great deal of time engaged in this activity and will use a dog to help out. Being veterans of losing pheasants to the wandering instinct (especially in the days before we fully understood the benefits of 'intelligent' cover cropping), we firmly believe in the benefits of 'dogging-in'. If you decide to get involved in this please do make sure your dog knows what it's supposed to be doing. There's absolutely no point letting a deadly enthusiast off the lead that will simply blunder about and send the birds off to the next county. Take the whole thing quietly and move the birds back gently. We prefer to use a shepherding rather than a utility or gundog breed for this. Either way it's certainly not obligatory to have a gundog to do the job.

The idea, of course, is to round up the poults and herd them back towards the release pens and drive areas before they become impossibly lost. Sadly, in our case, whilst we have the right breed for the job we don't have the right dog. Sam, our Australian Shepherd, has been around for as long as our children can remember. Desperately willing but now stone deaf and barely able to walk, he viciously guards the front doorstep whilst pretending to be asleep. But the poults still need dogging-in so the next in line is the husband – a bit rickety like the dog, but gets around well enough and is probably a bit more biddable. The point being that, despite cover crops, you have to be prepared to put in some legwork to bring them back in from time to time.

5

BROODY HENS AND STOCK MAINTENANCE

'MEANWHILE, BACK AT THE RANCH'

Broody Hens

Our decision to try using chickens as broodies came mainly as a result of speaking to several gamekeepers. They considered it to be a successful way of producing chicks and especially partridges, which are particularly vulnerable in the early stages of their development. Rather than buy stock in, a kind neighbour donated a mixed bag of eggs to help us get started. Varying in sizes and in shape, the only thing we could be sure of was that they belonged to a motley assortment of chickens. We set them in the incubator to see what happened. Sod's Law – we produced seventeen cockerels and six hens but decided that this was enough to begin with.

We did not take a gung-ho attitude to the process, but we did start off with a bit of a 'just how difficult can this be?' mindset. We nurtured the chicks to chickenhood with no difficulties at all. Extremely robust birds in every way, they developed rapidly. Oddly enough, some of them grew into the most unusual creatures with great plumage apart from completely bare, skinny necks. I later discovered that they were a hybrid turken. I'm sure their mothers love them, but they do look a bit odd. Anyway, on the basis that looks aren't everything and we needed the other end to do the work, we set about making our broody hen buildings.

We built two chicken coops with spacious runs and divided our hens up with three to a cockerel. We did this as a 'security measure' in case we had a predator invasion or outbreak of illness in one. The remaining cockerels were either given other jobs to do, or eaten.

Hybrid turken.

I admit that, with hindsight, we probably did not research the broody process sufficiently. Our thoughts were geared more to rearing the birds competently. Full of naïve enthusiasm, we thought we could just line up the girls in nest boxes full of game bird eggs and leave them to get on with it – or something like that. Not so! Apparently the whole process has to be conducted on their terms, which is why you should forget any precisely timed hatch schedule if chickens are your sole incubators of choice.

So, rather than list the things that can go wrong, the rest of this chapter addresses how to choose the right breed, and reproduce the right conditions, to encourage your hens to become broody. And when they deign to do so, how to manage the sitting period successfully.

The Broody Breeds

Some chickens actually have the brooding instinct bred out of them because it gets in the way of the laying capacity. However, there are others that do have a greater natural tendency towards broodiness. The well-known bantams, a smaller version of the standard-size chicken, and silkie crosses, are amongst the most popular. A purebred silkie is characterised by its feathers and skin, which is dark purple. The feathers are not webbed together as in ordinary chickens but are very soft, having the same consistency as the fluff around a hen's backside. Whilst they appear very delicate, with fluffy topknots and bodies, they are hardy, resilient birds provided they have the usual shelter and food afforded to other breeds. According to the Poultry Club of Great Britain either the silkie cross wyandotte or silkie cross Sussex hybrids produce the best broodies.

If you want to be a little more adventurous you might try the following less common breeds. The cochin is a heavily feathered hefty bird, fairly rotund and rounded in appearance. They are very large, with cockerels weighing up to 5.8kg (12¾lb). They make excellent broodies because they are docile and possess a maternal nature. But, there are some inherent health problems not helped by the fact that they are pretty lazy, and like the silkies don't seem to fly much. This, coupled with the very heavy plumage, renders them susceptible to metabolic and pulmonary problems. Given their size it is arguable that they may not be suitable for partridge eggs, but their reputation for excellent brooding tendencies is attractive so may be worthy of further research.

The smaller, similar version of this is the pekin. This again is a very placid bird. Feathered in a similar way to the cochin they also have the reputation of being excellent broodies and great mums. As with the cochins they don't need much space but they do require grass to forage. If this is not available they can also develop health problems. As already stated there are no guarantees, but these may be worth a look.

Hen Coops

Proper housing is important for hens and the principles are no different from keeping game birds. There are many different designs on the market and organisations to take advice from, but here is a basic 'safe' guide if you choose not to free-range them.

You will need an enclosed coop and a run. As with game birds, hens cannot sweat to air-condition their bodies. Instead they pant to remove excess body heat. I mention this detail because unless you have direct experience of a 'hot' bird slightly extending its wings, and panting as though it's fit to burst, then it's easy to mistake the behaviour for some respiratory condition. Don't panic, it's just hot. To compensate for this you must site the unit in an area that affords both shade and sunny spots. Chickens like to dust bathe, so the perfect ground mix is grass and a spot with bare earth.

If you are constructing from scratch consider building a unit with a detachable run, the benefit being that you can move the birds to fresh ground from time to time. Fit a couple of simple pop-holes in the shed and periodically move the run from one site to the other. Ours are not designed in this way because we modified a couple of old sheds and attached permanent runs to them. To balance this we made the runs much larger than necessary and every now and again fence off one section to allow fresh grass to grow through. If you take this option 2 × 2m (6ft 6in square) is fine for a static enclosure accommodating up to six medium-sized hens. As always though – the bigger the better.

The coop needs to be airy, and contain both individual nesting boxes and group nest areas. If you want the hens to lay eggs to eat when they're not brooding, make sure natural or artificial light can penetrate the coop. This stimulates the pituitary gland that encourages the laying process. Nest boxes should be about 45cm (18in) square and can be made from many different materials including natural wood, wooden ply and plastic crates. We line these with a fine layer of dry soil then put wood shavings or straw on top. They should be positioned well off the ground. To access the eggs in the least invasive way possible we have hinged doors that open directly onto the nesting area. We can then remove the eggs with minimal disruption to the birds. This has added advantages for easy access when cleaning.

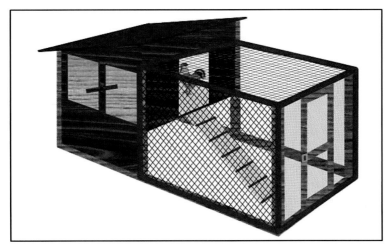

Hen coop with raised nest area.

The birds will need roosting space. A good rule of thumb is 30cm (1ft) per bird. Make sure you position them 1–1.50m (3ft 3in–5ft) off the ground. We used slim tree branches to make the perches, which seemed to go down well. Just make sure they are a suitable size and type for the hen to grip easily and don't position them directly above the nesting boxes, for obvious reasons.

Brooding

The next challenge is to get your hens to become broody. There genuinely appears to be no scientific method by which this can be done – it's hormonal. Poultry enthusiasts have endless suggestions as to what might work. Here is list of the more realistic options:

1. If you are feeding some kind of 'layer' pellet this should be changed. It is rich in calcium which the hen does not need at this stage, and may pass through her system too quickly, causing her to defecate on the eggs. Change this to a very simple diet of mixed corn or a chick granular formula which is useful preparation for the chicks.
2. Place a number of artificial eggs in the nest (starter clutch concept).
3. Same concept but leave real eggs in the nest, making sure you mark them in some way so you can tell the difference between the 'dummies' and the real ones. These can then be replaced as necessary.
4. Leave eggs laid by the hen in the nest to allow a natural build-up.
5. Make sure the 'nest' areas are dark and secluded. Hens favour a hidden spot that they instinctively feel is safe from predators.
6. Coop construction must be cosy and conducive to laying.

Poultry expert Mike Hatcher gives excellent advice with reference to creating an ideal site, which is reproduced here by kind permission of New Holland Publishers:

> The nest should be about the size of the one she usually lays in. Place a *turf of grass* in a box with the centre pushed down and a thin layer of hay [we prefer chopped up straw or wood shavings because of the moulds that can be present in hay] on top. It will help if you shape the bottom a little so that the centre is lower, to stop the eggs rolling into the corners. Some hen keepers choose to leave the nest open for the bird to come and go as it wishes. Others close the nest and pick the bird off once a day to let it feed and drink and to empty its bowels.

N.B. *Avoid using fine sawdust; this can be ingested by the chicks and get stuck in their throats.*

If you think your hen has become broody but can't decide don't worry, the signs are clear. First she will remain constantly on her nest and appear to flatten herself in an attempt to fill the space. If she is still sitting after about three days you can be cautiously optimistic. Sometimes her comb will gradually change colour from red to pink (especially noticeable with bantams). If you approach a broody hen she will fluff up her feathers, eye you beadily with great suspicion and often make growling-type noises. She can also assume a trance-like expression.

Your litmus test is to place your hand either near or gently under her bottom to establish if there are any eggs beneath. If this is met by either a raucous squawk coupled with a swipe from a wing, or a vicious flurry of pecks, then you're probably on to a winner. Additionally broody hens have a tendency to produce runny droppings that smell. As Mike Hatcher explains, 'The hen will only leave the nest around once per day, so a lot of waste has to be evacuated in one go.'

If you leave her in the 'team' coop you run the real risk, as has happened to us twice (slow learners), that the other hens will compete to lay on the same nest and either disrupt the broody or, worse still, scatter the clutch, breaking a few eggs in the process. So, if you think a hen is broody the best advice is to remove her from her nest. Do this last thing in the evening when she is least disrupted and place her on a new nest, which should be situated in a dark, secure, quiet place. Leave her for a couple of days to make sure she is still content to sit before gently introducing your game bird eggs and removing the others. Again, do this last thing. The number you can incubate will, of course, depend on the size of the hen but between eight and ten pheasant eggs to a bantam is normal.

Keep an eye on your broody. All being well she will look after herself as well as her clutch, but not always. If you find that she won't budge at all you will need to help out. She's also likely to have a good dust bath. If you decide to intervene you must stick to a regular time. As with game birds these are routine-driven creatures and you do should not disrupt her natural rhythms by random visits, which may put her off. Since hens are not deeply bright you also need to make sure that when she returns to the nest she is sitting *on* her clutch rather than *next to it*. If it's the latter you'll know what to do.

For the final day or two before the due hatch date, leave well alone. She will be aware of the chicks 'pipping' and will want to remain with them. Hatching may take up to three days, so resist the temptation to intervene until the fourth whereupon you can cautiously check any unhatched eggs for signs of fertility. The unhatched eggs will normally rattle if you shake them, but do so gently because this may well risk an unwelcome explosion!

Once your clutch has hatched the hen will normally be an excellent, highly protective mother, but don't be over-hasty in returning the new family to the rest of the flock. The natural pecking order is such that the chicks may become bullied or killed – the term 'pecking order' is not an accident. The hen should remain with her brood until the chicks have become feathered.

Don't forget to make sure you have made all the relevant adaptations to feeders and drinkers so that once the chicks begin to forage they don't accidentally drown in a moat-like water base, or starve because they can't reach the food containers (as outlined in Chapter 3). You should also check the pen carefully for any chick-sized holes in the mesh or base.

Meanwhile…

The May/June/July period is very busy. At this point you will be involved in some serious multi-tasking activities. Our time is taken up with a combination of raising chicks in the brooders and hardening off others outside. We're managing several in the rearing pens, and keepering those already transferred to the release pens. We're also nurturing broodies. All this, while sticking to the routine maintenance chores in the breeding and poultry pens. It takes a certain amount of discipline and dedication to handle properly.

There will be periods, particularly during bad weather, when you would rather be doing something completely different that has nothing at all to do with livestock. But it is also a good time to step back and review. Give yourself the odd mental pat on the back. After all, to have reared and maintained groups of game birds and chickens at various periods of development to this stage is no mean feat. Add to these your game crop management and predator control activities, and you should be justifiably proud of your achievements thus far.

By late June, you will have picked all the eggs you need for the season. If you're not sure when to stop, just remind yourself of the release process and start of the shooting season, then work backwards. If you continue to incubate your eggs into July they will not hatch until the end of the month or August. This means they will not be ready for release until September at the earliest. The logistics just don't work. It's too late if you want to begin shooting during the conventional open dates, let alone the proximity to winter. So for obvious reasons an eye on the calendar and attention to your deadlines are a must.

In our experience hen pheasants won't stop when you want them to. They can continue to produce eggs into October. The only thing that seems to halt a contented layer is being able to brood on a clutch, or the change in weather to colder conditions. So there will be a surplus. Make no mistake, these leftover eggs will need to be picked up regularly and used in some other way or, if absolutely necessary, disposed of. We made the basic error in our first year of leaving all surplus pheasant eggs in the breeding pen. Our naïve idea was that it would encourage them to 'go broody', make a nest and create a clutch. This was a particularly bad notion as many of these eggs were strewn around the pen. It was probably a little optimistic to hope that they would somehow be gathered together to form a nest. Added to that was the real risk of attracting vermin into the pen.

So the eggs have to be collected. Nobody likes an exploding egg, especially our country-born turned city slickette daughter who was despatched to undertake an egg clearance job. She approached the task with her usual verve and enthusiasm, looking suspiciously neat for the job. Rather unfortunately, whilst in a prone position retrieving a couple from a tricky spot, she shovelled them a little too gustily. The eggs exploded with an innocuous 'pop', splattering seriously old rotten egg all over her face. To her credit, almost undaunted, she ignored stifled guffaws from the rest of us and continued valiantly. We approached the job very differently thereafter.

So, if you want to leave some eggs for the broody females, make some kind of mark on them so you can identify the date each was laid. Gather only the fresh random eggs and form some kind of rough clutch to get them started.

The Nesting Game Bird

Pheasants

As I have explained, pheasant hens will not always follow through and incubate a clutch. They're probably completely put off the whole idea, having had their eggs mysteriously removed on a regular basis leaving the creation of a clutch seeming impossible. However, if you want to let Nature take its course, they certainly can, and if this happens you may find it attracts others to lay in a similar spot. It's just the same as chickens. We have gone to great lengths, and provided sufficient faux nesting areas for all the birds, but this can be for naught.

The more tricky situation arises when you have several hen pheasants (this has never happened with our partridges), all focusing on the same nest. This presents two challenges. The first is where the other hens clearly believe the

Pheasant hen brooding on her nest.

Partridge hen brooding on her nest.

nesting hen is 'team sitter', and so they all lay on the *same* clutch. In this completely unnatural situation the clutch eventually becomes unmanageable for one bird. The other problem arises where they join in, leaving three or so hen pheasants all trying to incubate the same clutch which again progressively grows larger and larger.

Inevitably in either case if you do not intervene several of the eggs get rolled to the outside of the nest, go cold and fail, and others are broken. The problem is compounded because, unless marked, it is impossible to know which to remove. Certainly the hens don't seem to know, which in turn makes effective incubation difficult.

We have not always found this easy to manage with a contented group of breeding stock hens, but we now follow these rules to help things out:

1. Always date-label eggs you decide to leave.
2. If you don't want the hens to nest, clear away the eggs at least twice daily.
3. If you have a brooding hen that you are happy to let sit you need to modify her environment. Create a suitable nesting area that is difficult for more than one hen to access. We have achieved this very simply by banking up the sides around the established nest using either planks of wood or upturned crates. We then mark the eggs. We also make sure the whole site is loosely surrounded by extra branches. Fully protected from inclement weather and other users, she has a good chance of going 'full term' with her clutch.

We normally do this last thing in the evening when, as with the chickens, they are more docile. The hen may behave protectively by hissing and pecking and will probably rush off the nest in a fit of hysterics. Don't worry – that's a positive sign of broodiness and, apart from sustaining the odd bruise, we've never had a problem with her returning. All being well she will return very rapidly after you have finished interfering. If all of this causes you to become 'head-shy', wait until she comes off the nest to feed and bathe, and do it then. The true broody is an extremely determined creature.

4. We have also successfully moved both the hen and her clutch. Our adult breeding pen is designed with two areas that can be split off to form three separate enclosures. Normally we keep the whole area open, but it is handy to be able to close off the smaller sections for this situation. The move process can be a bit trickier, but once done and the hen is safely isolated with her clutch, after a short bout of nerves she usually returns to the nest. If you're going to try this do so when the bird is calmer e.g. at nightfall.

5. The hatch and early care stage is exactly the same as with the chickens.

6. This comment goes for both species: have an eye on the calendar. There is no point allowing the hen to produce a magnificent clutch of chicks just in time for Christmas – it's simply not fair on the chicks, they'll end up as canapés.

Partridges

As I have already noted, both grey and red-leg partridges can be more difficult to rear than pheasants. If you have harvested sufficient eggs for your incubators and, as with the pheasants, view any natural births as a bonus then that's fine. Thereafter it is quite possible to use broody hens and, even more reliably, the parents themselves. Specific advice regarding grey partridges is given by Dr Mike Swan, Head of Education with the GWCT:

> Greys reared like red-legs and released in coveys made up of just young birds have very poor survival, and virtually always fail to breed. So, it is usually much better to try to allow their parents to rear them. This means fairly large pens where they can settle as pairs and go broody. You can also rear in huts, just as with red-legs, and then make up coveys by fostering groups of young greys to ex-laying pairs.

Breeding pen construction for the 'greys' is also described by the GWCT and includes the following tips: 'Keep pairs in 20 × 10ft [6 × 3m] pens, divided into two halves and connected with a pop-hole or door. One half of the pen should consist of short grass, a feeder, grit and a drinker, whereas the other half should be old tussocky grass including brashings for nesting cover.' Further detailed information is available in the GWCT 'Releasing Document'. It is an excellent resource, and well worth a read.

The system we now follow for our 'red-legs' is a little different. They are kept in two groups of around ten (numbers fluctuate from year to year), and in much larger pens of 10 × 14m (33 × 46ft). They pair off naturally in the spring and create their own nests in one of the sheltered parts of the pen. We have never had significant problems with aggressive males, but this may be because the pens are so big.

Red-leg partridge hens are less chaotic than pheasants with their nesting arrangements, but sometimes they will build two nests. They will lay eggs in each, and the clutches will be incubated by both the paired male and female. You will certainly know if the hen has become broody because she is usually fiercely protective of her clutch. I can bear testament to this having been chased around a pen on several occasions. Once you are certain they are brooding the best thing to do is make sure they have adequate shelter, with food and water readily available, and then leave them to it.

Breeding pen for red-leg partridges.

As I explained earlier, using bantam or silkie hen broodies can also be very effective. If you decide to use a broody hen to incubate the eggs make sure you allow the partridge to build up a number before removing some. The same applies if you are picking for the incubating machines. You can replace them with artificial equivalents so the hen believes the nest is still intact, but this doesn't always work, so don't be greedy and take too many in one go. Partridges definitely react differently from hen pheasants in this respect, so if you do take the whole lot you are likely to put the hen off nesting all together.

There is another technique you can try. That is to remove and artificially incubate the eggs right up to the last two or three days before the hatch date and then place them under a broody hen for the birth, the logic being the benefits derived from using artificial incubators and the reduced human workload in the rearing process. But, it does require one to have broodies 'on-tap' at exactly the right moment which, sadly, we have never had, and thus we have never actually done this.

A further tactic follows the broody hen principle. This again involves removing the majority of the partridge eggs and replacing them with artificial ones. Put the eggs in the incubator whilst letting the partridge sit on the faux clutch. After twenty-one days do a swap so she ends up hatching her own eggs. The theory is that you have a more reliable hatch percentage and this method also allows you to hold back a few eggs to minimise failures. We have never tried this either and are not wholly convinced that our canny lot would be hoodwinked by an almost completely artificial clutch.

The fact remains that, if you are trying any of the techniques that involve removing eggs, it's best to at least try to give the 'faux' version a go.

Handling the Surplus

After the 'city-girl' incident we now remove genuinely surplus eggs at least once each day. This coincides nicely with the routine husbandry visits, so does not become yet another over-burdensome task.

The thought of throwing away perfectly good eggs seems ridiculous. We absolutely hate waste, so rather than chucking them out we give away and eat as many as possible. Excellent in taste, game birds' eggs can be used in exactly the same way as those of the chickens. The principle is the same as with game meat. We always make full use of the birds we shoot, not to mention the rabbits we control, or the odd squirrel! And as for the veggie trimmings, there's plenty available in the woods and hedgerows; and sometimes even a little surplus from the cover crops too.

I discussed the importance of using these 'excesses' with Rachel Green. Rachel is a chef, farmer and food campaigner who is passionate about game food. She has been closely involved with the Countryside Alliance Foundation Game-to-Eat initiative, and has a wealth of experience in creating recipes that make full and sensitive use of the fruits of the countryside, and British game. She said:

> I love cooking when I'm at home with my family and there's nothing better than sitting down to dinner and enjoying natural produce such as pheasant, rabbit or partridge. It's immensely satisfying to be able to cook with game you've caught yourself. My family have been farming since 1650 and throughout the generations we've been taught that food and specifically food from the land is a precious commodity; if you're going to go out and shoot something at least have the respect to eat it as well, there's nothing quite like 'bringing home the bacon, or should I say pheasant!'

To illustrate her point Rachel has kindly suggested one of her recipes that we might try.

Roast pheasant with braised turnips, chestnuts and apples.

Serves: 4

Ingredients:

2 young pheasants, oven ready
1 small onion peeled and halved
4 pieces streaky bacon
2 tbsp butter
Sea salt & black pepper

500g turnips peeled and chunked
2 cored red apples, thickly sliced
1tbsp of fresh sage, chopped
150g peeled chestnuts

100ml cider
50ml vegetable stock
2 bay leaves
50g butter
Sea salt & black pepper

Instructions

Pre-heat the oven to 190C/Gas mark 5.

Season the birds inside and out. Place a small knob of butter inside each pheasant and half the onion in each pheasant. Cover the breast of each bird with 2 rashers of streaky bacon. Place in a roasting tin and cook for 40 minutes, basting from time to time. Remove the bacon and return the birds back to the oven for a final 10 minutes to brown the breast. Make a thin gravy with the roasting juices. Keep warm.

Meanwhile, melt the butter in a heavy based casserole. Add the turnips and saute in the pan for 3–4 minutes. Then add the chestnuts and cook for a further 2 minutes. Season well then add the cider, vegetable stock and bay leaves and cook in the oven with a lid on under the pheasants for 20 minutes. Remove the lid, and add the apple and sage and mix in well. Cook without the lid for a further 8–10 minutes.

Then drain the turnip mixture and put on a serving platter with the pheasants at the side.

Add the remaining vegetable juices to the gravy and serve.

Roast pheasant with braised turnips, chestnuts and apples. *(Photograph courtesy Michael Powell)*

Rachel also uses pheasant and duck eggs which, in her opinion, are perfect for her potato and wild mushroom cakes with duck/pheasant eggs and spring herb butter.

Managing the surplus in this way just makes sense. But make sure you don't inadvertently poison your diners. Don't forget where you are with your treatment programme. There are several medicines that require egg withdrawal for around one month after the treatment has been given, so make sure you check the manufacturer's instructions before tucking in. This, of course, is also true of your released birds. Chemical treatments must be withdrawn before the shooting season begins.

Finally, we do use a number of unwanted eggs as bait. For example DOC (Department of Conservation) in New Zealand traps lend themselves to this. Small predators like stoats and weasels in particular love eggs. When setting the bait, use gloves to eliminate human scent. You may not succeed the first few times so be prepared to check; and if you use the sort of trap where the egg is cracked open, change the bait daily.

Despite your best efforts you may still be left with a glut of eggs. Avoid creating an environmental disaster when you dispose of them. Bin them responsibly by bagging them up, preferably in vermin-proof containers.

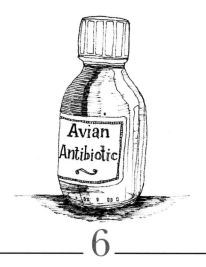

6

HYGIENE AND MANAGING DISEASE

'DON'T FORGET THE DUSTER'

Interim Clean-up

As the rearing season wears on, by July your incubators will be empty and most of your birds hardening off and in the rearing pens. If we have time during this period we start to 'deep clean' the machines and incubator building.

As with the nurseries you will be surprised at how much dust and fragments of downy fluff accumulate during the hatching process in spite of regular cleaning. We are rigorous about this routine and will wash down all the walls using a proprietary aviary disinfectant and insecticide. The incubator fan and machine parts will be de-clogged using an air compressor and then, for the washable parts, we use the same aviary cleaning products, in both spray and basic liquid form. Some of these cleaners are extremely strong so you should get used to wearing security glasses to guard against splashes. Gloves are a must, as are dust masks, which help guard against hopelessly bunged up sinuses and the 'albeit very rare' risk of an invasion from zoonotic bugs. Wearing overalls is also sensible if you want to avoid the 'tie-dye' effect on your clothes.

Once you've thoroughly cleaned up you can safely stow the machines away in dust covers for the winter, with the confidence that they will only need a superficial clean at the beginning of next season.

Disease

When you are involved in rearing game birds you may be faced with having to deal with injury or an outbreak of disease. Taking any form of proactive approach towards critical care is obviously useful. We have two small 4m (13ft) square pens which we use to isolate any of the stock that needs individual attention. For example, if we find a poult has been injured in some way, we will immediately remove it to the 'hospital wing'. Here (assuming we don't need to call in the vet), it can be cleaned up and treated with antiseptic spray on the affected area, given a couple of days to heal completely, and then returned to the flock. Happily, instances of this kind are very few and far between, but we prefer to be prepared. It also avoids the development of unwanted bullying behaviour from the others.

Similarly, with a bird that is 'off-colour', we will remove it, and treat the symptoms accordingly. Clearly this process does not work with mass outbreaks of disease, or illness associated with parasitic infestation on a large scale. It is also likely to be a hopelessly unrealistic approach if you are involved in rearing thousands of game birds. But if every one of your stock counts, then it can be an efficient way of returning a bird to full health. Equally, it safeguards the rest of the flock, by quickly removing the affected bird to a safe place. It goes without saying that you must be absolutely scrupulous with your hygiene routine, and be prepared to change the site of the 'hospital' pens to fresh ground if necessary.

I have allocated a fair amount of space to this subject because it is of vital importance to be able to recognise some of the symptoms of disease in game birds, and thereafter act quickly. It is equally important to understand the (often very basic) ways to give yourself a fighting chance of avoiding them.

Fortunately, as is often the case in the game bird world, there is a great deal of good information available in the public domain. It's just a matter of finding it. I have sought extensive advice from Richard Byas, a recognised specialist in this field. Richard is a highly experienced veterinary surgeon who has written many articles on the subject and continues his research today. I am extremely grateful for his help.

Early Experiences

Unfortunately, in spite of the most stringent bio-security, it is still possible for your birds to develop an infectious disease, or become ill as a result of parasitic infestation. Regrettably, both scenarios are conceivable in a situation where birds are raised intensively using artificial means.

To the untrained eye, early signs of health problems are not always clearly evident. But, once it becomes obvious that some of your flock are ill, you need to be ready to deal with it efficiently. The sad truth is that, despite the fact both pheasants and partridges appear to be robust, they can fail very rapidly and are extremely good at dying. This is especially so when young, before they have had the chance to build up natural immunities. So your vigilance and continued observations are essential.

We learnt the value of prompt action the hard way. During a routine feed round late in the autumn of our first year I noticed one of our hen pheasants behaving a little strangely. Just occasionally she seemed to twitch her head and make a kind of sneeze action. She followed this by opening her beak once or twice as if trying to swallow something – and then nothing. At that stage we didn't worry and assumed it was probably a fragment of food that had become temporarily stuck in her throat. Over the next few days her behaviour continued and became worse. Changing our diagnosis, we naïvely wondered whether this could be a respiratory bug because other than the 'sneeze' and odd erratic twitch of head, the bird seemed to be fine. I decided at that stage, rather than rushing out to buy armfuls of drugs, we should wait and see if the problem sorted itself out.

Unfortunately her condition deteriorated, and others began to exhibit the same symptoms. Becoming a little anxious, I called the vet for some telephone advice. She explained that the bird might be suffering from either a mild infection, or that it was more likely to be a case of infestation by *Syngamus trachea* (gapeworm), a parasitic nematode. The condition colloquially known as 'gapes' is produced by the gapeworm. The vet prescribed a general antibiotic that would treat both problems, together with a specific treatment for the parasite.

Before we even had a chance to start the medication we found two dead birds. On the vet's advice, I took one of them in to be autopsied, the result of which proved to be money well spent. The diagnosis was correct, the bird did have an infestation of the gapeworm in her gullet, which had basically suffocated and killed her.

We immediately began a focused treatment programme which arrested the spread, but lost a sixth of our stock in that pen. Oh, and let's not forget that, because of our slow reaction time, we also effectively sowed the seeds of

HYGIENE AND MANAGING DISEASE

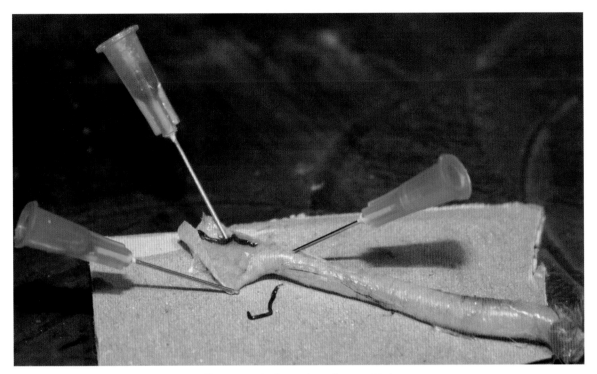

Gapeworms in the trachea of a game bird.

gapeworm in that particular enclosure for some years to come. We later learnt that despite treating the soil with lime, closing the pen up for twelve months and re-seeding it, it probably wouldn't do that much good. The parasites can live for up to five years in earthworms, which of course, in turn, are eaten by the birds, so could continue to present a health hazard in the future. Quite honestly, if you are faced with a similar problem, if possible you're better advised to re-site the pen.

I should also add (and any experienced gamekeeper will tell you), that the gapeworm is one of the most common parasites that can attack both pheasants and partridges, so some forewarning of what to expect is extremely handy.

The moral of this sorry story is, act on your instincts – if you think something is not right with the birds don't panic, but do move quickly. Delaying can be deadly and costly, especially if you have low stock levels and not much space. If you don't know what the problem is, don't be embarrassed to call in the vet. We all know how much veterinary visits cost, but you need to treat them as an investment rather than expenditure. And if you have mystery deaths that you simply can't explain, consider the post-mortem option. It gets you to the core of the problem quickly, and avoids wasting precious time ruminating over hypothetical illnesses.

Proactive Practices

There is a whole range of common-sense proactive measures that can minimise the frequency of disease, and help manage the outcomes. But first, it's helpful to have an understanding of the risk areas. Richard Byas provides clear advice about the potential problems that we need to be aware of:

> Obviously the first priority is to try to prevent disease being present in the pens to infect the birds. Most diseases are not present in newly hatched chicks so are carried to the birds from outside sources. Possible sources of disease include:

1. Sick birds already present.
2. Wild game birds present on the land.
3. Introduction of sick birds.
4. Parasites already present.
5. Rats, mice, flies, cats and dogs (spreading disease on their feet, etc.).
6. Use of dirty or contaminated equipment.
7. The land on which you pen the birds.
8. Other shoots locally with significant numbers of birds.
9. People visiting other sites with game birds present.
10. Visitors (including advisers and veterinary surgeons) visiting the property.
11. Dirty water, food and bedding materials.

Reducing the Effects of Parasites Present

Even with strict biosecurity, it is often not possible to totally prevent disease organisms affecting birds.

The most significant difference between naturally reared and artificially reared birds is the stocking density. At higher stocking (*basically birds raised in numbers in captivity*), when parasites are excreted or birds have an infection, they are more likely to infect another bird than at lower stocking densities (*in a natural setting*), and so numbers can multiply rapidly.

In the wild, infections tend to be self-limiting as the birds develop some degree of immunity or tolerance. However, when infections are overwhelming or if a bird's immune system is depressed then clinical disease may develop. In these situations a number of strategies are necessary:

1. Avoid overcrowding of the birds in the pens. Reducing bird numbers can significantly reduce the level of parasites and other diseases present.
2. Improve the immune status of the birds where possible. Stress reduces the effectiveness of the immune system which diminishes the birds' abilities to resist infection.
3. Reduce the level of infections present and reduce the birds' contact with infective organisms.

It is worth mentioning, with reference to point 2 above, that the word 'stress' in this context actually means anything that adversely affects the birds' immune system, or anything that reduces the birds' ability to fight disease thus making it more likely to become ill. Richard Byas qualifies this:

1. Physical environment factors that will increase stress:
   - Poor temperature control, chilling or overheating of chicks.
   - Day/night temperature variations.
   - Draughts.
   - Lighting intensity and length of daylight.
   - Rain.

2. Psychological factors that will increase stress:
   - Constant predator threat.
   - Overstocking – birds lower down the pecking order will be stressed.

3. Diseases, some of which specifically affect the immune system, for example:
   - Mycoplasmosis.
   - Immune system overload in contaminated environments.
   - Overwhelming infections – overstocking.

4. Nutritional
   - Standard diets should provide the requirements for maintenance and growth of the birds. In game birds higher levels of vitamins A and E enhance immune competence.

In tackling the risk of disease outbreak, apart from following a strict hygiene regime, you can take other practical measures. For example, using medicines can be highly effective. One word of warning though – do *not* launch into this with wild enthusiasm, assuming a cocktail of various products will be the ultimate cure-all. Take proper advice from your vet, and work with them to produce a suitable medication programme that will take the birds through each stage of their development.

Drugs can be administered through food, water or injection. Some are available on the market and others are by veterinary prescription only. Just make sure you buy them from an approved source, and read the instructions very carefully because dosage and storage advice can differ significantly. **N.B.** *If you need them, take reading glasses when dosing – minute type can be tricky, and you don't want to make mistakes.*

Medicinal products are intended to help guard your flock against the most common diseases and parasite infestations, and can be used for both preventive and curative purposes. They may be given periodically and often in place of the usual game bird feed, or pure drinking water. But not everyone likes chemicals. As an alternative, or to balance the chemically based products, there is also a wide range of complementary medicines, which are available in both granular and liquid form. These are typically aimed at improving the immune system of the bird, thereby reducing the incidence of disease. They are normally used as a preventive, although many will be used as part of a treatment programme, which is our preferred practice.

Of course if you do experience an occurrence of disease then you may need to react by altering your normal routine but, unless you are highly experienced, you are best advised to consult your vet first.

Common diseases

There are several diseases that commonly affect both game birds and chickens. I include chickens here because, if you are using broodies to incubate eggs, then you will also need to have an awareness of their health vulnerabilities. I have listed some of the main ones and shown which species is/are susceptible – including us.

One of the thoughts that may occur to you, as you scan this chart, will be that many of the symptoms exhibited are similar. To compound matters, your eye will also be drawn to the treatments table which often states 'none'. It all makes for depressing reading on the face of it, but there's no need to become overly worried. You do need to be made aware of some of the health problems that can affect game birds, but keep it in perspective and be comforted by the overall messages. Most of these diseases can be avoided by adopting efficient husbandry practices, together with an appropriate diet that often includes medicated foodstuffs. Support these by avoiding overcrowding your stock, and always diligently observe the birds' behaviour. If health problems do emerge, seek expert advice if necessary to help you spot problems early and nip them in the bud.

HYGIENE AND MANAGING DISEASE

Diseases that may affect game birds; causes and treatment.

BACTERIA

Name	Usual Transmission	Common Symptoms	Prognosis	Prevention	Treatment	Notes
Erysipelas	oral and contact	septicaemia; lameness	high mortality	vaccination	antibiotic	can be caught by humans (zoonotic)
Mycoplasma gallisepticum	air, oral, contact	flu-like symptoms; bulging eyes	can be fatal	hygiene and healthy diet	antibiotic	recovered birds can be carriers
Salmonellosis	oral and contact	diarrhoea, feeble, weight-loss	treatable	hygiene and healthy diet	antibiotic	can be caught by humans (zoonotic)
Tuberculosis (avian)	oral	diarrhoea, feeble, weight-loss	high mortality	hygiene and healthy diet	none	
Omphalitis (yolk sac infection)	unsanitary conditions	swollen abdomen, diarrhoea	high mortality	hygiene in hatchery	antibiotic	chick disease via unhealed navel
Colibacillosis	air and oral	feeble; poor growth rates	often fatal in young	hygiene and healthy diet	antibiotic	E-coli infection
Bumblefoot	bacteria through wound	swelling of foot	deformity	good housing conditions	none fully effective	vitamin A may help

FUNGUS

Name	Usual Transmission	Common Symptoms	Prognosis	Prevention	Treatment	Notes
Aspergillosis	inhalation	gasping, weak, thirsty	often fatal	clean litter and feed	none	raking bedding can worsen spore level
Candidiasis	Oral mouldy food	difficulty swallowing, weight loss	poor development	hygiene	anti-fungal	can be fatal in young birds

PARASITES

Name	Usual Transmission	Common Symptoms	Prognosis	Prevention	Treatment	Notes
lice, mites and fleas	habitat	feeble	anaemia; rarely fatal	difficult	acaricides	may require several applications
Syngamus Trachea	oral (soil and faeces)	coughing and snicking	often fatal	hygiene; pen rotation	flubenvet	common name: gapeworm
Capillaria	oral, (ground infestation)	feeble, diarrhoea, weight-loss	may be fatal in young	avoid overcrowding	flubendazole	common name: hair-worms
Coccidiosis	oral	feeble, diarrhoea, weight-loss	may be fatal	hygiene; early medication	coccidiostat	intestinal parasite
Hexamita/Spironucleus	faeces	feeble, weight-loss, diarrhoea	may be fatal	hygiene and healthy diet	none directly	treatments focus on bacterial enteritis symptoms and rehydration salts
Histomoniasis	oral (earthworms)	listless, stop flying, diarrhoea	may be fatal	hygiene; pen rotation	none directly	treatments focus on bacterial enteritis symptoms and rehydration salts
Trichomoniasis	oral and contact	diarrhoea, feeble, weight-loss	may be fatal	hygiene and healthy diet	none directly	treatments focus on bacterial enteritis symptoms and rehydration salts

VIRUS

Name	Usual Transmission	Common Symptoms	Prognosis	Prevention	Treatment	Notes
Fowl Pox	insect bites	lesions on skin; mouth infection	may be fatal	vaccination	none	
Newcastle disease	air, oral, contact	imbalance, feeble, neck-swelling	high mortality likely	vaccination	none	notifiable, common name: fowl pest
Corona-nephrosis	air, faeces	loss of appetite; withdrawn	up to 50% mortality	hygiene and healthy diet	none	
Avian Flu	contact, oral, faeces	flu-like; reduced laying	often fatal	bio security	none	notifiable, common name: fowl plague
Rotavirus	oral	diarrhoea, feeble	often fatal	hygiene and healthy diet	antibiotic	chick disease; treatments focus on secondary infection
Marble Spleen	air	difficulty breathing	often fatal	hygiene and healthy diet	none	birds are immune after recovery

OTHER

Name	Usual Transmission	Common Symptoms	Prognosis	Prevention	Treatment	Notes
Enteritis (Diarrhoea)	various	feeble; diarrhoea is often brown/bloody	treatable	hygiene and healthy diet	varies depending on source	avoid stress eg overcrowding

General advice

1. Wash all picked eggs in an approved product before setting. Take care to wash them in a solution that is warmer than the egg. This prevents contamination being sucked into the egg via the shell.

2. Appropriate housing and heating are required to provide steady temperature control in the first few weeks, thereby reducing stress.

3. Palatable fresh water and high-quality feed should always be available.

4. Dietary changes should be gradual and include preventive medicines (chemical and complementary), given in food and in liquid form. All these are designed to boost the birds' immune systems and develop a natural balance.

5. Store all medicines and vitamin products appropriately, normally in a cool, dark place. Adhere to the 'use by' dates – there's not much point religiously treating your flock with an ineffective product.

6. Don't overstock pens. Use common sense; understock rather than overstock. Also remember that, if the weather is poor, poults may need to be kept in the pens longer than planned – and they will keep growing!

7. Ensure that adult birds are not present on the rearing fields and in release pens, because they are a potential source of infection.

8. Disease is easily transferred on boots and hands. We site plastic containers (cat litter tray size), filled with avian disinfectant products, outside each pen and dip our boots in after every visit. We also use a pump-action bottle of bacterial hand wash (the type you see in hospitals), that dries instantly on your hands. Slot it into a holder on the gate next to the latch of each pen so you can't miss it. It soon becomes second nature to have a squirt after you close up.

9. At each stage of their development, take all measures to reduce the birds' stress.

10. Reduce the birds' direct contact with the potentially contaminated areas around feeders and drinkers by taking them off the ground or providing a clean base. A wide variety of options are available.

11. Parasites love warm, damp conditions. Immediately remove any leaking drinkers and never empty dirty drinking water onto the ground within the pens. Any accidental spillages will have taught you that birds think this is a great game and will be instantly attracted to it – a very bad idea.

12. There are some disinfectant products that can be safely applied whilst the birds are present. These products can be useful in reducing parasite/insect numbers. Ask your vet or Defra.

13. If you are experiencing a period of wet weather, make sure any bedding (e.g. wood shavings, straw) is removed as soon as possible afterwards and replaced with clean, dry materials.

14. Periodically during and at the end of the season take scrupulous care over cleaning all equipment that has been used and store it appropriately. Burn, or carefully dispose of, any kit that is broken or useless. Use only avian approved cleaning products; a bottle of household liquid just isn't strong enough. Always wear appropriate safety gear.

15. Before you start the next season give thought to bringing your vet in to check over your kit and pens, and provide you with a treatment plan ahead of time. We do this every year in case a new product that we don't know about has come on the market. It also provides us with our preventive 'meds. and vits.' programme for the new season.

16. Make sure you only use approved medicines and don't combine products without advice.

17. When preparing medicines for the drinkers always add the medicine to the water rather than the other way around – this avoids the whole mixture overflowing in a mass of bubbles and causing you to have to start again.

18. Make records of disease outbreaks and occurrences as you go along. Depending on the size of your shoot, this may, in any case, be a legal requirement. Either way, it is handy to remind yourself what happened and how to take future preventive action.

19. Make sure you withdraw any medicines that are not fit for human consumption, well before the start of the season. Note the dates of withdrawal. Similarly, be careful about the medicines you can and cannot safely give to your chickens if you intend to eat the eggs and surplus stock.

7

GAME CROPS AND LEGAL COVER

'BE PREPARED'

If you are a veteran shot you will be very familiar with the thrill of pheasants being flushed high from cover crops in front of you as they fly towards woodland safety, or perhaps partridges as they explode from a dense pre-prepared thicket. In the excitement of the moment it's easy to misjudge the amount of preparation that has to go into creating this setting.

In addition to predator control, one of the big challenges after releasing your own game birds is being able to 'hold' them. Ideally you will want them to naturalise in zones that make suitable drive areas for the Guns. Game crops are a proven method of achieving this, but you need to know what you're doing.

There's no point creating a wonderfully impenetrable jungle or vast area of barren meadowland if neither you, your beaters, nor your dog can find the birds or flush them. Failure at this stage can risk wasting all the weeks and months of hard work taken to rear the birds to release age. Therefore you need to understand how to develop your territory to its fullest potential. This is especially the case if you're new to the land, and it must happen months before the season opens.

Game crops are used specifically for your game birds and not for farming purposes. If you are shooting on rented or grace and favour terrain you need to make this clear to the owner. Better still, draw up a contract. Gentleman's agreements can often lack integrity in the heat of a debate about previously accepted site usage. Normally you will pay a specific rent for using the land in this way, so it is worth exploring the types of financial assistance that may be available to help fund the project. These are briefly discussed at the end of this chapter.

Analysis and Preparation of the Land

In planning your game crops you need to take a methodical approach to the process. These are some of the questions you should be asking:

- What do I want the crops for? For example, spring nesting, cover for released birds, feeding and holding, driving on shoot days?
- What will my soil type and climate allow me to grow?
- Is there a persistent weed challenge that may cause me to modify my choice?
- What funding and time limitations will cause me to modify my choice?
- Do I have to modify my crop choices to meet the needs of my game birds?

Jobs for crops may include providing a wind break on a bleak exposed site, or a nesting enticement that encourages permanent residence in suitable woodland. You can also create lead-in strips linking to cover or release areas, and even extend the size of existing natural cover, or perhaps provide increased spring nesting cover. There are several possibilities which will cause you to choose different sites with different crops. Unsurprisingly, making the right matches is considered by many to be a tricky skill to acquire. So, if you don't get it right first time, it's quite normal to take a 'suck it and see' approach, especially on a new shoot, and change your tactic the following season.

Once your site decisions are made you are getting closer to making some selections of suitable crops. This is where an understanding of your soil type comes in. If you don't have the knowledge to undertake these jobs give serious thought to bringing in an expert. Most of the top-quality crop specialists can provide a full soil-testing service. This includes measuring pH (acidity/alkalinity of the soil), level of liming required to balance any acidity for optimum crop growing, available phosphorus, potassium and magnesium levels, and major nutrient and trace element levels. This should be done in

February and you may well have to pay for the service but, as with other essential jobs, treat it as an investment. It's a total waste of effort and money planting cover crops that you *think* will do the job if you're not sure. You may end up watching them fail to germinate or die off rapidly.

If you do decide to have your soil analysed make sure you take samples from several sites. This should be recommended by your expert, so don't misjudge it as a further money-extraction ploy; accept the idea gracefully. Your soil composition is likely to vary so it's best to check each area designated as possible cover crop zones.

Defra can also provide specialist contacts from whom to obtain suitable advice and you might also talk to your local farmer. Ours was grilled on several occasions and gave us some very valuable advice on what would and would not work – we also got a regular load of manure to boot, which was a great bonus. But beware; make sure you're both singing from the same hymn sheet. Don't forget that farmers have a very different planting and crop yield agenda from you, so interpret the advice according to your own needs.

Climate

This plays an important part in determining whether or not a particular crop is naturally likely to succeed. Even small geographical areas can experience significant variations in temperature and precipitation. In fact these can sometimes be extreme enough to force us to make specific crop choices and compromises because of the special needs of a location. It's true that most brassicas and kale tend not to have geographical limits in the UK, but many of the others genuinely do, especially the further north you go. To reinforce this point, consider the different weather patterns for those living in the north of Scotland compared to Kent, or those experienced by people living in the west of Ireland compared to Lincolnshire, then look at the crops that are typically grown in each of these regions. It's simple; some plants definitely flourish more successfully in one climate than another. Of course topography and soil types are also influential, but the weather is nevertheless a key contributory factor.

Unfortunately, there are occasions where common sense gets hopelessly overridden by a burning ambition to plant a particular crop (often recommended by a semi-green-fingered colleague). This would be a perfect idea except that it won't grow in your area! Fortunately, help is at hand for the misguided amongst us. The research by top seed companies into developing hybrid varieties that will withstand significant climatic variations is producing excellent results. But this still means that you need to understand which seeds have the greatest chance of thriving, the result being that you may have to compromise on your former bright ideas by choosing alternatives. If you live in an area with a testing climate all of this becomes much more important. In taking your decision, whatever else, guard against buying-in cheap products. You are far better to go the whole hog and bring in good-quality seeds with the right treatments: these will have a much better chance of establishing.

Armed with this information you can then make some informed buying decisions. Consider it a kind of custom-fit programme.

What Crops, When and How?

I have already explained that different plants serve different purposes so, to give you more of an idea, I have provided a 'starter' list which describes some of the most popular game crops that are used in Britain and Ireland. In compiling this I have taken advice from two of the UK's leading specialist seed suppliers, Limagrain and Kings Game Cover and Conservation Crops, who both operate at the cutting edge of the industry, and Countrywide Farmers plc, the leading supplier of products and services to the rural community. The combination of their expert advice may persuade even the same-seed diehards amongst you to consider a new option or two.

Tried and Tested

Arguably the most popular classic game cover crops are maize and kale, followed closely by sorghum and triticale. Millet and quinoa as complementary crops are also favourites. Each of these single-sown species are typically used for winter holding, and share common benefits because of their reputation for early vigour and standing power.

Maize ranges in size from 1.2 to 2m (4ft–6ft 6in). It is an excellent source of food and cover and, depending on the type, can be sown as late as June. Successful establishment is particularly important for this plant in achieving optimum crop growth. Maize will not tolerate cold, compacted seedbeds and if planted in these conditions will often struggle to reach its full potential. This can prove to be a major challenge, especially since in recent years we have seen frosts continuing well past traditional maize drilling dates. For these reasons it is a difficult crop to develop in Scotland, so check for suitable varieties. If overall weather trends continue, with frosts lasting longer, it is certainly preferable to consider planting later. Crops drilled during the warmer weather, e.g. in late May and early June, often demonstrate exceptional vigour and strength, with many sown earlier requiring a boost from foliar feeds. So there's usually no need to panic; it may pay to be flexible

Although many growers plant maize as a single crop, both Kings and Limagrain suggest that it does well when planted alongside strips of kale, sorghum or a wild bird seed mixture. In southern regions Kings recommend under-sowing maize with a red and white millet mix whilst in northern areas they suggest that spring triticale appears to work well, both providing additional feed and warming the crop floor.

Sorghum is ideal for driving, making superb flushing points within maize, and adds warmth when sown adjacent to maize. The larger variety is also an excellent windbreak. Its versatility and ease of management are well known, particularly when compared with some of the establishment problems associated with kale crops and bugs.

We have used a combination of sorghum and maize very successfully for a number of years; the sorghum in particular stands up well to the damage caused by roaming deer. We plant them adjacent to one another in 200 × 30m (220 × 33 yard) strips. We don't drill until mid-June when the soil is much

Sorghum.

warmer, and so far the crops have remained very healthy, providing our birds with a strong winter hardy cover and feed crop.

Although not without problems, kale can be a great economical choice. It provides two years of cover, and there are several varieties that possess excellent germination and vigour potential. As Countrywide Farmers explain: 'The main attribute of kale is its ability to hold birds right through the winter whilst its clean stems allow birds to travel freely. It produces a leafy canopy which also provides warmth.' As with all brassicas it does need fertiliser, so would not be ideal for set-aside sites where you cannot use artificial nutrients. Because of its winter hardiness, it is also suitable for driving game. In the second year it will provide cover and feed for game and farmland birds. Caledonian kale is a variety that can be sown on brassica-sick sites (sites that have the club root fungal disease, which is caused by the *Plasmodiophora brassicae* plant parasite, in the soil).

However, one downside with kale is its reputation for being difficult to establish owing to the challenges from flea-beetle (discussed later), pigeons and slugs, as well as restricted weed control opportunities. However, progress has been made in terms of managing these problems. Several seed merchants now offer a range of varieties which are available with seed treatments that boost early plant establishment.

Caledonian kale. *(Photograph courtesy Limagrain)*

Millet can be grown by itself and also makes a good mixer. Its large seed head provides lots of good-quality feed, and it is consequently widely recognised for its value to game and farmland birds. It will grow well on most soils and in the right conditions can be easily and rapidly established whether broadcast or drilled. It does prefer warmer soil, so don't drill too early. Millet makes a good companion crop to maize or sorghum by providing increased warmth and shelter.

Another plant you might consider is quinoa. For those of you unfamiliar with this plant, quinoa is not technically a grain; it is actually the seed of a leafy plant. Its relatives include, beets and Swiss chard. It is capable of growing in poor soil, dry climates and even mountain altitudes, but thrives best in well-drained soil. It combines well with kale. If fertilised properly, quinoa provides an excellent source of high-protein food as well as cover.

Perennials

Kings believes that perennial crops have a vital role to play on all land managed for game and farmland wildlife. Many have the potential to provide cover for several years, and are especially useful in areas that are difficult to access for regular cultivation. They are also excellent if you do not have much management time or manpower. This is of particular interest to the 'DIY' shoot where most of the jobs are done on a part-time basis. The inclusion of small areas of biennials or perennials can definitely reduce the workload during the drilling and establishment period.

A further advantage relates to the increasingly variable weather in spring and early summer. When conditions prevent you from preparing and planting when you would like, this is another good reason to use some areas for perennial crops.

Additionally, and certainly a real plus for us, perennials provide refuge at lean times of the year. They can offer valuable nesting cover, and an insect-rich zone for young broods to forage in. This is particularly important during the early spring period, when most annual cover crops have been flailed and ploughed in, and for the partridge rearing community introducing perennials is a definite advantage, we thoroughly recommend it. Examples include artichokes, miscanthus and reed canary grass. Each is used for a different purpose, which includes cover, energy fodder, windbreaks and nesting crops. But, if planting canary grass take care; it does have a distinct tendency to march off all by itself, so you should drill in wide rows to make sure it doesn't get out of control.

Mix and Match

Mixing crops can be extremely effective because of the potential for their versatility. It really depends on what you want the crops to do, and the challenges presented by your soil. For example, a mixture of triticale and chicory works well in tandem. The triticale acts as a nursery crop to the developing chicory, whilst providing valuable game bird food. Meanwhile, because of its deep roots, chicory is very drought-tolerant, and the combination of these crops provides excellent feed and game cover. If your passion is partridges, Limagrain suggests that a mixture of linseed, kale, mustard and sandoval quinoa will do the trick in keeping them happy. On the other hand, a simple mixture of sunflowers and maize adds a splash of colour to the crop, and also acts as a valuable food source for wild birds at the same time.

To complement this, Kings' 'Brood Rearer' mix containing birdsfoot trefoil, linseed, sainfoin, triticale, vetch and wheat offers several handy combinations. The mix of triticale, wheat and linseed creates a safe canopy, allowing broods to forage, safely hidden from avian predators, whilst the sainfoin and vetch act as an insect-attracting lower level cover, replacing the weeds normally found in a headland. It is worth noting in particular the other values of triticale: a wheat/rye hybrid that retains seeds throughout winter, it is an excellent source of nutritious food. A further benefit is that it does well on poor soils and will tolerate relatively low pH levels (i.e. acidic soils).

Brood Rearer mix. *(Photograph courtesy Kings)*

Back at the pens there is even a solution to worn soil bases or shady areas in rearing and release pens. Both Kings and Limagrain offer specific mixes sharing some common ingredients that include amenity perennial ryegrass and creeping red fescue. The beauty of these is their low maintenance and suitability for both laying pens and rearing fields. The formulas establish rapidly, and are designed to be used for both poults and adult birds.

The message here is that, whichever seed supplier you use, there should be a rich choice of options from which to make your selection. If you find you are not being given sufficient options, you are probably at the wrong supplier.

Catch Crops

Catch cropping can be a very helpful addition to game cover crops. These versatile plants are generally fast growing, and can either be developed simultaneously with, or in advance of, the main crop to be cultivated on a particular site. They can also be used to provide a replacement for an existing failed cover, or create a new crop after a cereal harvest. However, their success is very much weather dependent and to reach their full potential they will need moisture. If the seed is sown prior to a wet period, then quick establishment will result in a strong crop. In drier seasons growth may well be delayed until late autumn, so timing is crucial.

Crops sown in the autumn to be harvested before the main sowing/planting season in spring are not normally considered as catch crops. But, although often slow to mature, they do satisfy the basic definition of using ground left empty between crops. Late-season sowing generally avoids the peak of flea-beetle attack, and also allows a good weed control opportunity. They can also provide a secondary use as a forage crop after the shooting season. Seed can be drilled or broadcast into standing crops or directly onto stubble after harvest.

Catch crops can also be sown to prevent minerals being flushed away from the soil, an example of this being millet. Planting this crop offers the potential to keep certain minerals not attached to the humus–clay connection (such as carbon and other positively charged elements), in the soil for years. Other examples of successful catch crops include mustard, phacelia and radish.

Organic or Not?

If you decide that growing organic crops is the thing for you then, once again, the top crop companies have the solution. After all, with over 600,000 hectares (nearly 1.5 million acres) of organic production in the UK, it's not surprising that interest in organic cover crops is increasing. But it's not easy, especially if you want to make what amounts to a significant change at the same time as rearing your own game birds. To switch to an organic method of farming takes time and commitment, and may be unrealistic if you are renting land unless your farmer shares the same goals. The experts will tell you that, initially, the land ideally needs to be given over to an organic conversion period of three years. It is then often sown with a crop like red clover which improves soil fertility, and slowly removes the residues from the previous conventional system.

Take care when buying in your seed; select only bona fide companies who will advise on a range of organically produced seeds that can be used to produce covers with 'green manure' benefits. This involves the practice of ploughing or turning into the soil undercomposed green plant tissue for the purpose of the improving physical condition as well as the fertility of the soil.

Planning and preparatory work also needs to be well organised. For example where possible you need to apply farm yard manure. If this is not possible, once the crop has emerged, a foliar feed e.g. Algifol can be applied, which will supply valuable plant nutrients. This product is derived from seaweed and an algae extract; and is cleared for use on organic crops. It acts as a fertiliser that is absorbed directly through the leaf.

Cover For Your Birds

An appreciation of a few of the main plant options available is very useful, and will help you match crops with the needs of your shoot. Limagrain offers some ideas of the varieties of crop you might be looking at for both pheasants and partridges that serve different purposes. Final buying choices will, of course, involve taking into consideration the usual land, climate and resource constraints.

WINTER HOLDING

Purpose	Crops Used	Benefits
To hold and feed birds throughout the winter	Mixtures with feed and cover potential e.g. maize, millet, buckwheat, kale, linseed, triticale	Stops the birds wandering; reduces your feeding costs; can be driven on shoot days

DRIVING CROPS

Purpose	Crops Used	Benefits
Crops that are used on shoot days to drive game	Usually drilled, with easy access e.g. maize (check variety), kale	Makes a new drive; extends an existing drive

NESTING COVER

Purpose	Crops Used	Benefits
Crops designed to encourage nesting birds	Tussocky grasses with good structure e.g. sweet clover, reed canary grass, cocksfoot	Helps increase wild bird stocks

BROOD REARING

Purpose	Crops Used	Benefits
Crops designed for rearing young chicks	Open structure with easy access and insect-attractive e.g.WM1 (mix of three small-seed bearing crops plus kale and quinoa) e.g. Magnet (triticale, fodder radish, phacelia and linseed)	Attractive for insects and excellent for chicks, providing good cover and food source. Also helps increase wild bird stocks

RECOVERY

Purpose	Crops Used	Benefits
Crops designed to be sown later if initial crops have failed	Catch crops that can establish rapidly e.g. mustard, rape, fodder radish, stubble turnips	Summer sown game cover; patching up existing areas

Crops and their uses. *(Reproduced with the kind permission of Limagrain)*

Getting Prepared

As with any standard farm crop, a specialist game crop will only reach its potential if it is well organised and properly managed. As part of this, issues like the timing of sowing are extremely important and sometimes optimum timings will have to be compromised because of soil difficulties, adverse weather conditions or even lack of manpower. For example, if soils are difficult to work with in the spring, or you don't have enough time, it may be better to concentrate on a well-managed perennial like a canary grass. Similarly, if the game crop sites are of poor fertility, extra management will be needed to be certain of a good crop. If issues like this become a regular pattern of how things are on your shoot, then consider changing your choice of crops.

Rotation, Weed Control and Seedbed Preparation

That said, there are other things you can do to give yourself the best chance of growing a productive set of crops. As Countrywide Farmers notes:

One of the key elements to game cover crops reaching their full potential is rotation. It is essential to rotate your game cover crops to help reduce soil-borne diseases such as club root brassicas. A rotational system will also help to improve soil fertility and structure as each crop can benefit the soil in different ways, each requiring different trace elements. Crop rotation is essential where weeds and or diseases have become a persistent problem.

Weeds are potentially the biggest obstacle to successful crop establishment and, if you have a particular type that is a constant problem on your land, you are far better to choose crops that are not affected by it. However, half the battle is often identifying them, especially if the land is worked regularly. So talk to the farmer and find out which are the most troublesome. Then remove the risk by making 'safe' crop choices to avoid unnecessary extra work.

Weed problems can also be created with mixtures. Maximising value to other farmland birds by growing a range of different crops in different sites is a great idea, but beware when using mixtures because they can be hard to manage and sites can get very weedy. If you are short of management time but like the idea, if you have a large cover crop site a good alternative system is to grow several separate strips or blocks of different crops so that you can practise crop rotation within the overall site.

The fact that plots are cropped repeatedly, with mixtures often being used, practically guarantees a hefty build-up of weeds. We had a perfect example of this problem with a site that we planted with sunflowers. It had been planted up several times before with sorghum, triticale and wheat. We prepared the seedbed and set our seeds properly. Unfortunately other jobs got in the way and we completely overlooked the weed control side of things. By the time we realised our error it was too late. The sunflowers were too advanced, as were the weeds.

Failed sunflower crop.

We ended up with an impenetrable jungle that our poor game birds (not having been supplied with machetes) staggered around in and eventually gave up on, and practically no sunflowers. Great cover, but probably only for mice, and the area was later re-designated as an enormous hybrid beetle bank!

If you know you have a weed problem on a particular site the benchmark weed control method is the 'stale seedbed' technique. This involves creating a seedbed some weeks before seed is due to be sown. This preparation makes sure that any weed seeds that have been disturbed and brought to the soil surface during cultivation will have a chance to germinate, and can then be removed before the designated crop is sown. The technique can be utilised in early spring, when the weather is still too cold for proper seed germination. If a drilled crop can establish quickly in a weed-free environment, it has a much greater chance of beating off any later pressure from late weed growth, pests and disease. However, this technique does need rain for the seeds to germinate. So, if you are struggling with this process, go back to your supplier, or indeed the local farmer, who may be able to advise you of some other effective methods for weed management.

A well-prepared seedbed is essential for crop health and development. For most sites ploughing and rapid consolidation to conserve moisture will be the ideal start. For example, following the plough with a furrow press will not compact the soil but leave it ready for one more pass just before establishment. Moisture retention is especially crucial for crops sown in May when very dry conditions often occur. Seed requires good soil contact so small seeds, such as those of kale and quinoa, need fine, firm seedbeds. Larger seeds also need a reasonably fine seedbed for successful germination. This can be helped by rolling the ground after drilling, which in turn will help to retain valuable moisture.

Fertiliser use during the preparation stage will aid crop establishment. An application of farmyard manure will be very helpful in improving soil structure, increasing aeration and helping to retain moisture.

If you're not sure about this or any other techniques associated with preparing your seedbed you can always check with your seed supplier for further detailed planting advice and techniques.

Pests

The monitoring and control of plant pests is as important for successful game crop establishment as it is for arable and vegetable crops. The damage caused by pigeons, crows and rabbits can be significant, so setting deterrents such as flags, bangers and traps is time well spent. But just make sure you are not flushing your game birds into the next county in your endeavours!

Game birds themselves can also cause significant damage to establishing crops, particularly those adjacent to woodland areas. To try to combat this we place several feeders and drinkers around our cover areas. This serves a dual purpose; it helps hold the birds whilst diverting their attention from following drill lines and digging up emerging seedlings.

Slugs and other insects can also help ruin crops and, depending on the weather, are present each year in varying numbers. Cold, damp springs, for example, normally result in wet and claggy seedbeds which are perfect for slugs to take hold in. It's wise to keep a check on their progress, which may result in a sprinkling of slug pellets, but just make sure you use the correct variety that will not poison your game birds. You might also consider drilling slightly later and rolling the seedbed afterwards. This will also assist in preventing damage.

No doubt the single biggest issue facing brassica growers (and, according to the top seed suppliers, forcing many to move away from choosing kale) is the flea-beetle. It can be a huge problem. These are small beetles, just 2–3mm (approx. $\frac{1}{10}$in) long. They have large hind legs that enable them to leap off plants when disturbed. The adults feed on the leaves and the larvae on the roots. You'll know if you've got them because rounded holes are scalloped out of the upper leaf surface. Often these do not fully penetrate the leaf. The damaged areas then dry up and turn pale brown. Fortunately it is now possible to buy kale products with the option of various seed treatments that help prevent attack by flea-beetle, but don't lose sight of the fact that these treatments should be considered as an aid to controlling flea-beetle and not a replacement for good crop husbandry.

Trouble-shooting Dos and Don'ts

In summarising some of the typical problems experienced by crop growers, Limagrain offers some practical advice as shown in the adjacent chart.

ANIMAL DAMAGE	SOW
Rats and badgers are a problem.	**Chopper sorghum** Excellent controlled driving cover. Replaces maize. Sow in June. Height 120–150cm.
Rabbits eat and damage my crop.	**Labrador mixture** Two-year full season cover and feed. Ideal for use in cooler more exposed areas and ES suitable. Height 100cm. **MIX:** mustard; quinoa; keeper kale; spring triticale; linseed
	Spring triticale Resilient to rabbit damage. Sow in spring. Seed heads should stay late into winter. Height 90–100cm.
Deer are a problem in my maize crop	**Chopper sorghum** (see above)

DIFFICULT SITES	SOW
Thin soils with low pH	**Labrador mixture** (see above)
	Spring triticale (see above)
	Spaniel mixture: Excellent recovery potential. Sow in summer (into cereal stubble is OK). Drought tolerant. Used for patching. Height: 80–100cm. **MIX:** mustard; texsel greens, fodder radish, interval rape/kale
I can only sow in the autumn	**Spaniel mixture:** (see above)
	Magnet wild bird seed mixture: Sown Sept/Oct creates brood-rearing cover in spring. Seeds shed late summer onwards. **MIX:** Triticale; Fodder Radish; Phacelia; Linseed
I need a 'permanent' solution	**Springer mixture:** Very drought tolerant with excellent driving cover. Triticale provides food in year 1 and then in 2nd and 3rd years chicory flower at 150cm. **MIX:** perennial chicory and spring triticale
	Canary grass: Perennial grows on exposed areas and poor soil. Used for driving or nesting cover. Height 1.5m to 2m
I need to sow in a woodland area	**Buckwheat:** Thick bushy seed-producing plant. Seeds shed autumn. Very fast establishment with some weed-smothering properties. Can be sown in woodland glades. Height 70–120cm.
	Woodland glade mixture Cover up to 4 years. Useful for semi-shaded and open areas in woodland but not densely shaded areas. **MIX:** perennial chicory, sweet clover; buckwheat

ESTABLISHMENT PROBLEMS	SOW
I have problems establishing kale	**Caledonian kale (flea-beetle treated):** Bred with club root resistance. Can be continuously sown on brassica-sick sites. Winter hardy with good germination and vigour. Height 90–100cm
I need to control weeds with a herbicide	**Cocker mixture:** Butisan and herbicide tolerant. 2 year cover. Excellent feed for partridges and pheasants. Height 80–90cm. **MIX:** caledonian kale; mustard; fodder radish; linseed.
	Golden retriever mixture: Stomp Aqua tolerant. Excellent full-season cover and seed shed potential. Good for pheasant/partridge driving cover and for wild birds. Winter holding potential. Height 200–220cm. **MIX:** dwarf sorghum; dwarf sunflower; millet blend
	New pointer mixture: Stomp Aqua tolerant. Fantastic feed value. Excellent driving cover throughout the season. Height 150–200cm. **MIX:** maize; spring triticale; white millet; dwarf sorghum
My crop has failed – what can I sow?	**Spaniel mixture:** (see above)

Crop solutions. *(Reproduced with the kind permission of Limagrain)*

My own suggestions, in respect of some overall considerations, are as follows.

- Be prepared!
- Understand your game birds, climate and territory, and choose crops that match each.

- Be clear about the purpose of your crops.
- Plan to have year-round crop provision. Work towards one-third perennial cover, one-third game cover and one-third feed crops that qualify for the stewardship schemes.
- If financial and human resources are limited, consider contracting out the work, renting or borrowing equipment, and choosing low-maintenance crops.
- Do not stint on quality of seeds, especially if you live in an area with a challenging climate. You are buying seeds for their germination and vigour potential – it's worth paying more to get what you need.
- When buying your seed, ask what quality control procedures are in place. A high-quality supplier will sample and test seed on intake.
- Find out what the customer service arrangement is. Quality suppliers will operate a technical advice 'hotline' and will arrange on-site consultation visits if necessary. You may have to pay for it – but again, view it as an overall investment of time and funds.
- If you don't possess the expertise, bring in a professional who will test the soil for you, remembering to test more than one site.
- Get the seed on-site in good time so you are ready to plant when the conditions are right.
- Prepare the seed bed properly using recommended products to enhance its fertility.
- Look to plant the sites that are in reasonable order with the required crop. If some plots are not ready, wait and plant when conditions improve.
- Never underestimate the importance of weed control. If you have a particular type of weed that is ever-present on your land buy the crops that are least affected by it.
- Manage pests by understanding who the offenders are, and treating the problems accordingly. Take care not to terrify or poison your game birds accidentally in the process.
- Take serious account of the various government aid schemes; they are there to help.

The GWCT adds some special advice for rearing partridges:
- Construct beetle banks [a narrow strip planted with grasses or perennial plants that provides a home to insects and food, and cover for birds].

They are easy to establish and are positioned so that normal cultivation can continue.

- Manage the grass beside hedgerows so that there is always old dead grass from the previous year available for nesting.

- Keep hedges trimmed (preferably after the berry crop has finished) to under 1.8m (6ft) in height to avoid them being used as lookout posts by avian predators.

- Leave some bare soil between the hedge bank and the crops. It prevents weed infestation of the crop and provides a drying-out/dusting area for partridge broods.

- Make judicious use of field corners to create grassy nesting cover next to cereal crops with conservation headlands, or un-harvested cereal headland.

- Never spray out fence-rows with herbicides. Fence-rows are the only nesting habitats left in many areas.

- Leave stubbles as long as possible before ploughing. Stubbles following an under-sown crop are particularly valuable because they remain uncultivated through the spring within the ley.

- Try to avoid having livestock – especially sheep – to graze out and damage the base of hedges when adjacent fields are in grass. **N.B.** *Some light grazing every few years may be beneficial.*

Keep in touch with those you work with, whether it is the farm staff, your seed supplier or fellow game managers. Advice and support is out there if you ask for it.

'Setter' mixed crop used for winter holding cover. *Photograph courtesy Limagrain)*

Costs and Conservation

For those of us with one eye constantly on budget, and the other on doing the right thing, it is possible to save money and contribute to the conservation of our environment at the same time. This is through state-run schemes that are detailed and managed via government advisers on the natural environment.

There are several schemes available to farmers and land managers in the UK. However, this section follows advice supplied by Natural England and is specific to England. For other countries within the UK you will need to contact one of the following:

- Countryside Council for Wales
- Scottish Natural Heritage
- Northern Ireland Environment Agency

Although they do change, the following are some of the grant schemes currently offered by Natural England:

- Environmental Stewardship
- Catchment sensitive farming
- Energy Crops Scheme
- Soils for Profit (S4P) project
- Heritage Management Plans grants
- Nature Improvement Areas
- Local Nature Partnerships fund

One of the most significant is the Environmental Stewardship Scheme. As Kings explains:

> Environmental Stewardship now involves over 40,000 farmers who manage over 6 million hectares [over 14¾ million acres] of land in the schemes. With an annual budget of around £400 million, Environmental Stewardship has not only made an important contribution to the natural environment, but has also supported the rural economy through direct payments to farmers and improvements in farm assets.

Natural England describes the key elements to Environmental Stewardship:

Entry Level Stewardship (ELS) provides a straightforward approach to supporting the good stewardship of the countryside. This is done through simple and effective land management that goes beyond the Single Payment Scheme requirement to maintain land in good agricultural and environmental condition. It is open to all farmers and landowners.

Acceptance is determined by a simple points-per-hectare calculation across your whole farm. Providing you achieve your points target, meet the scheme conditions and agree to deliver the options you have chosen, you will automatically receive funding.

You earn points for the range of environmental management options you agree to provide over the course of the agreement. There are over 65 management options to choose from, suitable for most farm types. In *general* you must achieve a total of 30 points per hectare, and in return you will receive annual funding for it (a lower rate applies to larger parcels of land above the Moorland Line).

Organic Entry Level Stewardship (OELS) is the organic strand of ELS. It is geared to organic and organic/conventional mixed farming systems and is open to all farmers not receiving Organic Farming Scheme aid.

Uplands Entry Level Stewardship (Uplands ELS) was launched in February 2010 to support hill farmers with payments for environmental management. This strand of Environmental Stewardship succeeds the Hill Farm Allowance. It is open to all farmers with land in Severely Disadvantaged Areas, regardless of the size of the holding.

Higher Level Stewardship (HLS) involves more complex types of management and agreements are tailored to local circumstances.

I bring these schemes to your attention because you may find that you qualify for financial assistance. The best thing you can do is to contact Natural England, or its equivalent, for further advice.

Why Take Risks?

The simple answer is *don't!* The Countryside Alliance provides an excellent guide to conducting a full risk assessment on your land. Whilst it is geared more towards the professional, we used it as a helpful prompt. I have included extracts here to give you a flavour of the fuller advice. As you will see, the principles are applicable to us all, regardless of experience levels, or number of people involved in the shoot.

> Every shoot owner or manager who employs five or more people is required by law to *'prepare and as often as may be appropriate revise a written statement of his general policy with respect to the health and safety at work of his employees and the organisation and arrangements for the time being in force for carrying out that policy.'*
>
> This legal requirement extends not just to those shoots which employ a number of full- or part-time gamekeepers or other workers; it also encompasses those which employ people on a purely temporary basis, including beaters, loaders and pickers-up on shoot days. Moreover, such individuals may even be regarded as 'employees' if they receive only payment in kind. Thus there are few organised game shoots which are exempt from the need to prepare a written risk assessment.
>
> Recent legal decisions have made it all the more important for shoot managers to fulfil their responsibilities under the Health and Safety at Work Act and if they do not do so, then their shoot may not be considered 'lawful' under other legislation.
>
> **What is a Risk Assessment?**
> Assessment of risk goes beyond understanding and adhering to a code of practice. The law does not expect you to eliminate all risk, but you are required to protect the people working on your shoot as far as 'reasonably practicable'. A risk assessment is simply a careful examination of what, on your shoot, could cause harm to those who work there, so that you can weigh up whether you have taken enough precautions or should do more to prevent harm.
>
> The safety of others who share your workplace, such as farm, forestry or estate staff and contractors must also be considered, along with the safety of any members of the general public who may have access. Workers and others have a legal right to be protected from harm caused by a failure to take reasonable control measures.
>
> **Five Steps to Assessing and Managing Risk**
> Risk assessment is not rocket science. If you have been running and managing a shoot for more than a season then you will be well aware of the potential dangers involved, and you will almost certainly have taken steps to minimise them. A risk assessment is simply a formalising of this process.

Step 1 Identify the hazards and risks

A **hazard** is anything that may cause harm, such as firearms, vehicles, machinery, chemicals, electricity etc;
The **risk** is the chance, high or low, that somebody could be harmed by these and other hazards, together with an indication of how serious the harm could be. Walk around your shoot and ask yourself what might reasonably be expected to cause harm. If you are not the gamekeeper talk to those who are involved in the day-to-day running of your shoot. They may have noticed things which are not immediately obvious to you. The Health and Safety Executive (HSE) website provides useful guidance on identifying hazards associated with different industries. Check the operators' manuals, instruction books and data sheets of tools, equipment or chemicals which are used on the shoot. These will often assist you in assessing the extent by which such equipment or products may be considered hazardous.

Step 2 Who might be harmed and how?

For each hazard, think about **who might be harmed**. It is not necessary to list everyone by name but you should identify groups of people (e.g. 'loaders' or 'people walking on footpaths'). Remember that some groups may be at greater risk. e.g. new and young workers, elderly beaters, and those with disabilities. Extra thought may be needed for some hazards because visiting guns and guests, contractors, farm and forestry workers will not know the layout that well. Members of the public may also arrive, walking or riding on public rights of way, or simply trespassing!
If you share your workplace with farm or estate staff, then you will need to think about how your work affects them, as well as how their work affects you – talk to them, and don't forget to ask your colleagues if they can think of anyone you may have missed. Gamekeeping staff who are regularly employed either full or part time should be totally familiar with organisation and procedures. But remember they also deal with poaching, and may have responsibility for wider estate security. These tasks could incur particular risks.

Step 3 Evaluate the risks and decide on precautions

Having identified the hazards, you then have to decide **what to do** about them. The law requires you to do everything 'reasonably practicable' to protect people from harm. You can work this out for yourself, but the easiest way is to compare what you are doing with best practice. Ask yourself whether you can get rid of the hazard altogether If not, how can you control the risks so that harm is unlikely? When controlling risks, apply the principles below:
Try **a less risky option** (e.g. switch to using a less hazardous chemical);
Prevent access to the hazard (e.g. by storing cleaning agents securely);
Reduce exposure to the hazard (e.g. rearrange drives to prevent shot fallout over public rights of way);
Issue personal protective equipment (e.g. clothing, footwear, goggles etc to those using cutting equipment in the woods); and
Provide proper facilities (e.g. first aid and washing facilities for removal of contamination).
It is the shoot organiser's responsibility to ensure that everyone is provided with the means to look after his/her own safety and the safety of others on the shoot. The best way of doing this is through **structured briefings** These should include instruction about:
Low birds/Ground game/Location of pickers-up, stops etc. Loading and unloading procedures (e.g guns to be loaded upon arrival at pegs and unloaded when the whistle/horn is sounded). Guns to be in slips between drives. Special attention must be paid to new/young/inexperienced guns. If necessary, experienced loaders or instructors should be appointed to supervise novice guns. Communication can alleviate risk. Consider providing key personnel with two-way radios or mobile phones. Emergency signals and procedures should be clearly explained e.g. in the case of an incursion by hunt saboteurs. These will include making safe all shotguns and gathering at a pre-arranged location.

> **Step 4 Record your findings and implement them**
>
> **Write down** the results of your risk assessment in simple terms, for example 'Slipping on sleeper bridge: chicken wire fixed to bridge surface, staff instructed, annual check', or 'Falling from beaters wagon: tailgate latch fixed and regularly checked'. The law does not expect a risk assessment to be perfect, but it must be suitable and sufficient. You need to be able to show that:
> **A proper check was made;**
> **You asked who might be affected;**
> **You dealt with all the obvious significant hazards,** taking into account the number of people who could be involved;
> **The precautions were reasonable and the remaining risk is low,** and
> **You involved your staff or their representatives in the process**
> If you find that there are quite a lot of improvements that you could make, big and small, don't try to do everything at once.
> Arrange to train employees on the main risks that remain and how they are to be controlled.
> Make regular checks to ensure that control measures stay in place, and be clear on responsibilities. Agree who will lead on what action, and by when.
> Remember, **prioritise and tackle the most important things first**.

> **Step 5 Review your assessment and update if necessary**
>
> Changes to the organisation of your shoot or the introduction of new equipment, substances and procedures could lead to new hazards. You should therefore keep what you are doing under review. Every year or so **formally review** where you are to make sure you are still improving, or at least not sliding back.
> During the year, if there is a significant change, don't wait: check your risk assessment and where necessary, amend it. If possible, it is best to think about the risk assessment when you're planning your change.

The Health and Safety Executive's risk assessment checklist. *(HSE)*

The Health and Safety Executive recommend the five-point checklist reproduced here. If, having read the five steps, you still don't feel they are applicable to your shoot then just read on… you can probably add to the following Countryside Alliance (CA) list.

Examples of Hazards Associated with Rearing/Shooting:

Firearms

- Unsafe gun handling.
- Low birds.
- Ground game.
- Mixing different calibres of ammunition.

- Non-use of gunslips.
- Young/inexperienced Guns.
- Shotguns/firearms in unsafe condition.
- Use of rifles for deer/predator management.
- Inadequate gun security.
- Noise.

Vehicles

- Safety/suitability of Guns' and beaters' transport.
- Competence of drivers.
- Manoeuvring in parking areas/yards etc.
- Slow-moving shoot vehicles on public roads.
- Quad bikes/ATVs.
- Land rovers/4×4s.
- Use of trailers.
- Carriage of dogs.
- Carriage of guns and ammunition.

Shoot equipment

- Chainsaws.
- Brushcutters.
- Axes, knives.
- Winches, hoists.
- Incinerators.
- Welding equipment and all power tools.

Animals

- Gundog/terrier kennels.
- Ferret housing.
- Farm livestock.

- Inadequate hygiene.
- Dog bite.

Brooders

- Dust in hatcheries/brooders.
- Gas bottle storage.
- Gas connections and pipe work.
- Power – remember that all mains electricity powered equipment must be checked by a qualified electrician every twelve months.
- Lifting (feed for example).

Other

- Herbicides and other chemicals – remember all chemicals are potentially dangerous. The Control of Substances Hazardous to Health (COSHH) has issued strict guidelines regarding the storage and handling of harmful chemicals and the labelling that must accompany them.
- Working alone.
- Inadequate hygiene.
- Catering – if you use caterers for shoot lunches for example, make sure they have the appropriate food hygiene certificates.
- Structures such as bridges, stiles and high seats.
- Environmental e.g. ponds, rivers, rocks, extreme weather.

Regardless of the size of your shoot, there's a potential hazard for everyone so it makes sense to follow the advice. Both the CA and BASC have downloadable Risk Assessment forms which are helpful to use as templates.

Are Your Legal Arrangements in Order?

Allied with the assessment of risk is the importance of having the correct legal paperwork. Your people, guns and animals must be appropriately insured and, where necessary, you should have suitable leases in place. This is essential to protect your rights and hard work on land that is, and is not, your own. You should not take a casual approach to this. As with the whole of this chapter, it's all about creating the perfect habitat – safely, legally and responsibly. It just takes a little forethought and planning, and often some expert advice.

I asked Tom Devey, legal expert on rural affairs of MFG Solicitors LLP, to help with some of the more common 'legal' questions that may affect your shoot.

Syndicates and landowner legal relationship

QUESTION: What is the rationale for, and basics of drawing up a shooting lease?

ANSWER: Without a lease a syndicate can only be in occupation under licence and this licence or 'permission' can be withdrawn by the landowner at any time despite the fact that significant time and money may have been spent preparing for a season's shooting. It is therefore a good idea that a syndicate has a lease of the sporting rights for a determined period at a set rent. If a gamekeeper is selling days to a syndicate it is, for obvious reasons, vital that he is able to provide the shooting for which the syndicate members have paid. If he is not, he will be liable to them for the moneys he has received and the value of a well drafted shooting lease can't therefore be underestimated.

A shooting lease also benefits the landlord as he can plan ahead knowing that the syndicate are contractually bound to pay the rent specified by the lease and he can avoid the syndicate acquiring security of tenure under the Landlord and Tenant Act 1954 if the syndicate or gamekeeper also have use of a gamekeeper's cottage, game larder or other building.

Insurance

QUESTION: Is there a specific type of accident insurance cover required for the gamekeeper involved in the complete process: egg production and incubation through to release to wood and the shoot? *N.B. This will probably involve different people at different times and possibly the general public.*

ANSWER: Whilst business insurance may not be mandatory it is always advisable because if the gamekeeper fails to provide a well-organised and stocked shoot to whomever shooting has been sold then he could be liable and he should therefore seek suitable cover for such an eventuality. If, for example, all of the birds put down were stolen or killed so there were no birds available to shoot then the Guns would be within their contractual rights to insist that any money that they had paid to shoot should be returned and this could leave the gamekeeper in a difficult position. Bespoke insurance policies are available from specialist insurers.

With regard to insurance, whilst actually shooting it should be ensured that anyone coming onto the property with either a rifle or shotgun has the necessary personal shooting insurance through a recognised body such as BASC or The Countryside Alliance.

Risk assessment

QUESTION: Is there a legal obligation to conduct a formal risk assessment, or is this purely a matter of best practice?

ANSWER: Every shoot owner or manager who employs five or more people is required by law to 'prepare and as often as may be appropriate revise a written statement of his general policy with respect to the health and safety at work of his employees and the organisation and arrangements for the time being in force for carrying out that policy' (section 2(3) Health and Safety at Work Act 1974). *[This information from the Act mirrors the full advice given earlier in this chapter by the CA at the beginning of their guide to conducting a full risk assessment.]*

On the understanding that a syndicate is not a legal entity, the responsibility will lie with the individuals collectively on behalf of the syndicate as employer. This liability will be joint and several and a single member of the syndicate could therefore end up picking up the full cost of any financial penalty imposed.

Even if a shoot doesn't employ five or more people it is always advisable that, as a matter of best practise, a full risk assessment is prepared and it may also be a requirement for any insurance taken out.

Straying public

QUESTION: Up and down the country landowners have raised concerns in recent times that the public have been straying onto land when a shoot is taking place. How do public access rights affect the landowner and the syndicate's privilege to shoot?

ANSWER: It is not unusual to have a public right of way running through private land and this can affect the operation of a shoot significantly. For example, it is an offence to discharge a firearm within 50ft [approx. 15m] of a highway without lawful authority or excuse; if, as a result, a user of the highway is injured, interrupted or endangered. Whilst it is not an offence to discharge the firearm, provided no injury or disruption is caused, it is very unlikely that the 'lawful authority or excuse' would apply to sportsmen.

One point that should be made very clear is that whilst persons using a right of way have the right to pass along the piece of land, they do not have the right to loiter.

Private Land

For the avoidance of doubt, a trespass is committed by any person entering the land of another without permission to do so. Anyone who permits or invites another party to enter onto another's land without consent is trespassing. This even includes the sending of a dog onto neighbouring land to retrieve game. A Gun standing on his own land and who successfully shoots game which falls onto neighbouring land will commit trespass if he or the dog enters that land without authority. It is therefore well-advised that access to retrieve is discussed with any neighbouring landowners before a shoot takes place.

Whilst civil trespass is not a criminal matter for the police to deal with, aggravated trespass is, and the Criminal Justice and Public Order Act 1994 empowers the police to direct anybody committing an aggravated trespass to leave private land.

Countryside and Rights of Way Act 2000

If the land upon which a person wishes to shoot has been designated as 'access land' by the Countryside and Rights of Way Act 2000 there are further issues that will require consideration. Under section 22 of the Act, the landowner can make an application to exclude the public from land that has been designated as 'access land' for up to 28 days each year to allow shooting to take place.

However, section 42 of the Act provides for other statutory restrictions affecting public places, such as section 19 of the Firearms Act 1968, which makes it an offence to have a loaded gun in a public place without authority, to apply to 'access land'. It may therefore be that if land has been designated as 'access land' by the Act, and an application to exclude the public has not been made, then shooting on it has inadvertently been made illegal.

As ever, all landowners and syndicates must be 100 per cent clear on their boundaries, their rights and the rights of the public. Failure to understand those could make for a long and uncomfortable shooting season.

Types of shot

QUESTION: Are there any issues regarding the use of lead shot?

ANSWER: The use of lead shot is something that frequently comes up.

Although it is probably desirable that lead shot is not used at all, it is extremely important that lead shot is not used for the shooting of wildfowl as it may pollute the watercourses used by them. Furthermore if lead shot is used the landowner could be found guilty of polluting the watercourse and heavy penalties can then be levied by the relevant authorities.

The expert guidance kindly supplied by Tom exactly reflects the safety advice given by the CA and BASC, and is fully supported by practitioners. Similarly the information he offers regarding contracts and rights is vitally important for a shoot, regardless of size. I asked some experienced gamekeepers how they felt about the need for a small syndicate shoot that rents land to establish a legal agreement with the owner. There are no doubts at all – in every case the advice is: 'Don't go into a casual deal with the farmer or landowner without making sure a legally binding agreement is in place. It protects everyone concerned!'

8

PEST AND PREDATOR CONTROL

'IT'S A JUNGLE OUT THERE'

This chapter deals with both pests and predators. Pests do not directly kill game birds but they can indirectly cause the spread of disease, or death. Predators, on the other hand, are killers. Anyone owning livestock will certainly be familiar with the often wide variety of species that fall loosely into both categories.

Discussion on this topic generally revolves around problems experienced in the release pen and after, but there are many factors that can foreshorten the life of a game bird well before it goes to wood. Here are a couple of reminders:

Parasites are capable of causing higher mortality rates through disease than any of the predators we all typically complain about, and the birds haven't even left the rearing field. There's little point in getting obsessed about percentage losses to predators and rushing off to buy a gun if you don't also exercise due diligence in protecting your young birds from the ravages of parasites.

Rodents are bacterial disease carriers. Mice and rats will gnaw at anything, and mice in particular have a tendency to roam around the feed rooms, brooder areas and rearing fields. They will investigate the newcomers and explore ways of stealing food. They also seem to have a non-stop auto-defecation system, and will regularly litter food with their droppings. Since they are carriers of bacterial diseases the risk of the birds ingesting contaminated food cannot be ignored.

A further but less frequent problem is the rodent that has mysteriously died in the pen. It is quite feasible for poults to quickly dispose of the fleshy parts of the carcass, thus creating more potential health problems. Therefore all practical measures to control these small but significant pests should be taken.

Rodent Control

An obvious solution is to set suitable traps or use poisons, both of which I discuss later, but you can also engage the expertise of the house/farm cat, which can be highly effective especially when involved in protecting non-livestock zones like the food store (unless, of course, you have an unwilling participant). However, if you have one with more of the feline than fur ball about it (like ours), you should be fine. Of course the inevitable downside of using a cat for pest control is their liking for anything small with feathers. Sadly, they don't discriminate between the truly wild versus the raised to be wild. So, whilst we've never had a problem with the adult stock, we would never trust them with chicks.

To deter cats from patrolling the brooders we bought six solar-powered sonic emitters. They have an effective range of 9 × 6m (30 × 20ft). The type we use are small, inexpensive units that give off very high-pitched sound waves. Whilst being almost inaudible to the human ear the sound is highly penetrative and intolerable for animals. As to whether there is a lasting value we cannot be sure, but at the moment they certainly work on the cats, and appear to have helped suppress the rodent population. The units fix into the ground, which allows us to reposition them occasionally just to refresh the audio horror zone. And, since we have not experienced any problems with foxes here, we think they act as a useful deterrent for them too.

Poisoning Mammalian Pests

Poisons can only be used to kill rats, mice, moles and, under certain conditions, grey squirrels. Rats are particularly difficult to kill because, being cautious scavengers, they tend to nibble at several items of interest, pausing to make sure they do not become ill before continuing to feed. For rats and mice there are several compounds of poison (rodenticides) available, some of the most effective being the first-generation anticoagulants like warfarin, coumatetralyl and chlorophacinone. If they become resistant to these there is a more toxic list which includes difenacoum, bromadiolone and flocoumafen, the last of which must not be used outdoors. Odourless and tasteless, these substances can be very effective in controlling vermin because they are slow-acting poisons that do not give the animal any early-warning signs.

As with other predator control devices you need to locate the rodent tracks and runs, which are never that hard to find – just follow the trail of droppings to a hole. Lay bait in holes at intervals not too far apart. Rats in particular favour sheltered routes, so use these. Don't expect instant outcomes; because of the nature of the products it often takes up to three weeks to see positive results, so be patient and resist the temptation to change the bait every five minutes. These animals are extremely wily and too much human interference will only put them off.

If you lay bait in an exposed area it must be completely safe from accidental ingestion by non-target species including dogs, the heroic house cat and children. It is illegal to place poisons in the open or on an animal carcass, neither is it permissible to contaminate an egg because these could be taken by crows, birds of prey or domestic animals.

A final word of warning on minor mammals: if you are successful in trapping rats make sure you wear gloves when you handle the carcasses and dispose of them carefully. They are carriers of disease and parasites that are potentially harmful to humans and animals including leptospirosis (Weil's disease), toxoplasmosis and salmonellosis. If you need further information about using poisons for vermin control we found Natural England, the government's adviser on the natural environment, to be a very useful free resource.

Conventional Predators

Scientific Background to Predator Control

Dealing with the more usual mammalian and feathered predators is where good keeping really comes into its own. Predator control is absolutely fundamental to all game management as scientific research has repeatedly shown. The need has been demonstrated by the GWCT, who conducted a research project that examined the effectiveness of good keepering of grey partridges on the control of predators. Extracts from their findings are reproduced below with their kind permission.

Over more than 30 years, the Game & Wildlife Conservation Trust's work on the Sussex study area revealed the importance of predator control, where nest losses to predators were monitored over decades on farms with and without keepering. Then, during the 1980s, we conducted a controlled scientific experiment where we compared grey partridge population dynamics on similar areas of farmland that were and were not keepered. Over eight years on Salisbury Plain we showed that predator control:

- Increased the production of young birds.
- Increased numbers in August by 75% each year.
- Resulted, over three years, in a 3.5-fold increase in autumn populations.
- Increased breeding stock in spring by 35% each year.
- Resulted, over three years, in a 2.6-fold increase in breeding density.

In our current demonstration project at Royston, the effects of keepering and predator control are clear.

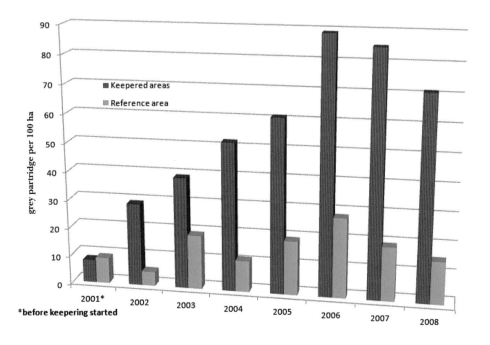

The effects of keepering and predator control. *(Courtesy the GWCT)*

Despite being aimed exclusively at the grey partridge the findings are similarly applicable to other types of game bird. I have included extracts from another piece of GWCT research that further illustrates the point.

In trying to establish the size of the predation problem on ground-nesting game birds the GWCT reviewed data from approximately 900 radio-tagged hen pheasants collected over a fifteen-year period on six different shooting estates. They analysed data from 450 nests and found that:

> 34% of pheasant clutches hatched successfully, 43% of nests were lost because of predators, a further 10% were abandoned completely and 5% were destroyed by farming operations. The remaining nests failed owing to other causes including flooding. 43% of nests were in woodland, 30% in arable fields, 13% in field margins, 7% in set-aside, 6% in grassland and 1% in other habitat types.

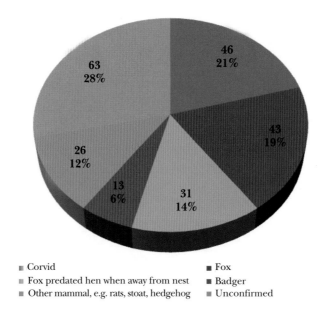

Causes of predation of nests. *(Courtesy the GWCT)*

Of the nests that were predated, foxes and corvids (mainly carrion crows) were the most important nest predators. Foxes accounted for 33% and corvids 21% of predated nests (see above). The estimates of fox and corvid predation are conservative as the nest predator could not be confirmed in a third of cases where the entire nest contents had been removed; something that foxes and corvids are known to do.

Sadly in our case a few dispiriting discoveries of ravaged nests amply supported the published results. Our difficulties arose not from a dislike of undertaking the task; it was more to do with trying to juggle too many balls in the air all at once, and consequently not spending enough time protecting our wild birds. Not surprisingly this is a common problem shared by many gamekeepers and will almost certainly become an issue for you, too, if you decide to raise game from scratch whilst encouraging reproduction in the wild. There are various 'tools' that help your control efforts; nevertheless, if you do have to prioritise your time for wild bird shoots you should concentrate predator control activities from February to July and for reared bird shoots from May to November.

So, after our second year, there was no avoiding the fact that we had to improve our rearing activities time management, and put in some shift-work on predator control. Being told by one experienced keeper to develop the mindset that every loss is one too many, and not being content with a 40 or 50 per cent recovery, certainly helped change our priorities.

The Law and Licences

But before you reach for your AK-47 and start blasting away at anything that doesn't resemble a game bird, you might want either to refresh your memory, or acquaint yourself with some of the things you can, cannot and ought not do. Since, in England alone, there are some forty pieces of legislation relating to wildlife management and licensing, there are many opportunities to fall by the legal wayside. And, don't be caught out by devolution either – laws often vary from country to country within the UK, so be on your guard for alterations in the detail.

It is your sole responsibility to know the law before you set any trap or snare, or shoot at any bird or animal. Ignorance does not constitute an acceptable defence so it is important to grasp the basics. As the BASC explain in their Code of Practice on the subject: 'In law there is a fundamental distinction between the protection afforded to birds and mammals:

- All birds are protected unless specifically excluded;
- Only specifically named mammals are protected.'

The Wildlife and Countryside Act 1981

This Act is one of the most important pieces of legislation that provides the source to these statements. It protects animals, plants and certain habitats in

Great Britain. It is amended through The Wildlife and Natural Environment (Scotland) Act 2011 (WANE), but does not apply to Northern Ireland, the Channel Islands or the Isle of Man (although these places have *similar* legislation in place). The Wildlife and Countryside Act was originally implemented to comply with the European Directive (2009/147/EC) on the conservation of wild birds and gives protection to native species (especially those under threat), controls the release of non-native species, enhances the protection of sites of special scientific interest (SSIs), and enhances the Rights of Way rules in the National Parks and Access to the Countryside Act 1949.

The Wildlife and Countryside Act works as a kind of statutory umbrella through the repeal, re-enactment and amendment of several other pieces of relevant legislation. It is split into four parts, and covers seventy-four sections and seventeen schedules. It is complex and lengthy so you may want to skip to the key bits, or alternatively take it to bed when there's absolutely nothing else to do. And if you can't manage any of the above here are some extracts that will give you an introductory flavour:

> If your business affects or involves wildlife you must not, unless you are licensed to do so (or you can rely on an exception or are shooting certain listed birds outside the close season), intentionally:
>
> - kill, injure or take any wild bird
> - take, damage or destroy any wild bird's nest while it is being used or built
> - take or destroy the eggs of any wild bird
>
> If you are within an area designated for special protection by Defra, you must comply with special bird protection rules for that area.
>
> You must not, unless you are licensed, intentionally or recklessly:
>
> - kill, injure, take, or trade in any wild animal listed on Schedule 5
> - damage or destroy or obstruct access to any structure or place used by a wild animal listed on Schedule 5, or disturb such an animal while it is occupying such a shelter or place.
>
> Unless you are licensed to do so, you must not kill or take wild birds or animals using:
>
> - any traps which are calculated to cause injury

- self-locking snares
- nets
- baited boards
- bird lime
- any poisonous or stupefying substance
- bows
- crossbows
- explosives
- automatic or semi-automatic weapons
- large shotguns
- night-vision equipment
- artificial lights or dazzling devices
- gas or chemical wetting agents
- live mammals or birds or sound recording decoys
- motor vehicles to chase wild birds or animals

(Certain exceptions apply under different sections. For example, Schedule (7) allows night shooting of ground game under certain conditions.)

Wildlife inspectors have powers to enter and inspect your premises in order to assess your licence application or to monitor and ensure that you are complying with the licence conditions.

If you occupy land within a site of special scientific interest, you must comply with requirements of Natural England.

If you do not comply with wild bird and animal protection rules you can be fined up to £5,000 and/or imprisoned for up to six months on summary conviction. On indictment this rises to two years and/or an unlimited fine.

General Licences in the UK

I have included these examples to illustrate how relevant the Wildlife and Countryside Act 1981 (and its amended versions) is for the gamekeeper. Frustratingly, several of the articles in the Act seem to hamper proactive attempts at predator control – however, note the repeated use of the words: 'unless you are licensed to do so'. Permission can be given under the terms of the General Licence. This is the mechanism that allows us to do the things that are otherwise prohibited in the governing Acts.

In describing the vagaries of this licence I have used information kindly provided by BASC and the CA. The BASC states:

> The first important fact to understand is that you cannot kill 'pest birds' at any time simply because they are 'pests'. This catches a lot of people out. You must understand the system and apply the rules correctly. What a General Licence does is to provide the exemptions that allow us to shoot and trap certain species.
>
> The licence is issued by Natural England in England, and the devolved governments in the rest of the UK. It provides a legal basis for people to carry out a range of activities relating to wildlife; actions that would otherwise be illegal under legislation contained within several Acts. This is renewed annually, which allows government bodies to alter the detail as required. This means you must check each year to find out whether there have been any changes. You can find this out from Natural England or BASC.
>
> General Licences authorise the legal control of certain species exclusively for the specific purposes stated on the licence. This is why, when you set out to control a pest bird you cannot say you are doing it for sport or because the bird is a pest. You must be acting in accordance with one of the licences and using an appropriate method of despatch. So make sure you have read them and don't get caught out if challenged. Licences broadly cover areas that are in conflict with people's interests e.g. air safety, damage to crops, public health risk and the conservation of other species. It is under a General Licence that most 'pest bird' species are controlled, such as pigeons, crows, etc. Control methods allowed under General Licence may also include shooting, the destruction of eggs and nests, and the use of Larsen traps, and multi-catch traps.
>
> It is a condition of the licences that you must be satisfied that non-lethal methods of resolving the problem are ineffective or impracticable. This is your personal judgement based on your experience and there is no obligation on you to have tried any such methods before you shoot.
>
> The preventative nature of many licences means you do not have to actually prove that the bird is doing damage when you shoot it; rather, the fact that it is capable of doing damage is sufficient justification. For example, magpies, through predation, can damage wild bird populations and therefore trapping them is a preventative measure to conserve wild birds.
>
> Under the terms of the General Licence an 'authorised person' is the landowner or occupier (or persons authorised by them, e.g. gamekeepers), upon their own land, or a person having written authority issued by local authorities or a person having written authority issued by the relevant statutory authority, regional water boards, river authorities or local fisheries committees.

You do not specifically need to apply for General Licences, but you are required by law to abide by their terms and conditions. It may sound rather obscure but, as you are highly likely to be involved in many of the activities outlined here, you need to understand the boundaries, and make sure that your actions comply fully with the conditions.

There are other pieces of legislation to bear in mind, some of which continue to catch the 'public eye' and attract media attention. The Hunting Act 2004 (England and Wales), which applies a great number of restrictions in England and Wales on how hunting with dogs is conducted, is one such example. This Act contains a number of relevant clauses. As the BASC explains:

- Up to two dogs can be used to stalk or flush mammals above ground to protect game, livestock, and food for livestock, crops, timber and fisheries. The user must have permission or own the land.
- A single terrier may be entered below ground provided the user is meeting the terms of the government-approved code of practice e.g. BASC.
- Using any number of dogs to hunt rabbits or rats is not illegal provided permission has been granted from the landowner.
- Hares cannot be hunted or flushed with more than two dogs. However more than two dogs can be used for the retrieval of shot hares.

The Protection of Wild Mammals (Scotland) Act 2002 is similar. There are exceptions, for example allowing certain types of hunting with a dog which is under control, as explained by the BASC:

- Stalking or flushing targeted wild mammals, for example hares, from cover above ground provided they are then shot.
- Flushing a fox or mink from below ground provided it is shot as soon as possible after it is flushed.
- Retrieving a hare which has been shot, or under specific conditions.
- Using a single dog to despatch fox cubs below ground and believed to be orphaned.
- Where a dog is being used in connection with the despatch of a pest species and the intention is to shoot that wild mammal once it emerges from cover or below ground, that person does not commit an offence if the dog kills the mammal in the course of the activity.
- Those excepted activities are conditional on persons involved being authorised and any killing being as humane as possible.

Predatory Behaviour and Control Options

Having scientifically established the relevance of controlling predators and understood something of the law, you need to be clear about the hunting behaviours of our most common predators. This allows you to determine the best ways of dealing with them. The adjacent 'most wanted' list is pretty standard, and you must acquaint yourself with those you can dispose of legally and those that are protected species. Where you live and the geography of your land will also have an effect on whether you typically have more or less of one type of predator or another.

					Available Control Processes			
		Problem	Protected	Risk (0–10)	Enclosure	Trap	Shoot	Deter
PEST/PREDATOR	Mice	EF/D	NO	3	#	x		x
	Rats	EF/D/KC	NO	3	#	x	AR	x
	Hedgehogs	EE	YES	1	#	x*		x
	Foxes	KB/KC	NO	10	x	x	R/S	x
	Corvids	EE/KC	NO	5	#	x	R*/S	x
	Raptors	KB/KC	YES	8	#			x
	Stoats	EE/KB/KC	NO*	8	x	x	AR	x
	Weasels	EE/KB/KC	NO*	8	x	x	AR	x
	Feral cats	KB/KC	NO	8	x	x	R/S?	x
	Mink	EE/KB/KC	NO	8	x	x	AR	x
	Marten	EE/KB/KC	YES	8	x	x*		x
	Grey squirrels	EE	NO	4	x	x	x	x
	Badgers	EE	YES	1	x	x*		x
	Owls	KB/KC	YES	5	#			x
	Polecats	EE/KB/KC	YES	8	x	x*	x	x
	Rabbits	EE/EF	NO	1	#	x	x	x
	Gulls	EE/KC	YES	5	#			x

EF=Eat food; EE=Eat eggs; D=Disease; KB=Kill birds; KC=Kill chicks
\# Only practical for early stages
NO* – YES in Ireland
x* must release afterwards
AR= Air Rifle; R=Rifle; S=Shotgun
R*/S Rifle only if prey is on the ground with safe backstop
R/S? Absolutely MUST confirm it is feral

Available control options.

If you're wondering why we have given foxes a risk factor of 10, speak to anyone who has had a fox raid on their chicken pen. The majority of other wild animals kill what they need to eat. Foxes kill everything within range, and then eat or carry home what they need. This can also apply to the weasel family, but foxes are bigger, and extremely good at it. From bitter experience there is nothing more depressing than finding 20 per cent of your poults destroyed by 'Mr Reynard' in just one overnight visit.

You will notice we have not included poisons in the table. Whilst I briefly described their usage earlier, we regard poisons as a bit of a loose cannon. There are experts who would disagree with this and may guarantee to get rid of rats without endangering the pet cat or dog, or other reasonably desirable species, such as small children, but we're not convinced. Poisons can only be used to destroy specified species under certain conditions. Obviously there are special techniques used but we continue to question how to guarantee the avoidance of poisoning non-target species.

Shooting is our preferred control option. It is target-specific, humane and, after initial purchases, relatively inexpensive. However, unless you are already an experienced and competent shot, there are a number of issues you will need to understand: firearms law; the choice of firearms for different quarry; the essential safety procedures; equipment preparation and practice; and the techniques that make the difference between a clean, humane, kill and an injured animal enduring a painful lingering death. These factors are dealt with in detail in the following section.

Firearms and Their Use

Firearms Law in the UK

This is covered by the Firearms Act 1968 and briefly states:

> If you intend to use firearms the legislation requires you to hold a firearms certificate. Section 1 of the Firearms Act 1968 makes it an offence for a person:
>
> **(a)** to have in his possession, or to purchase or acquire, a firearm to which this section applies without holding a firearm certificate in force at the time, or otherwise than as authorised by such a certificate;
>
> **(b)** to have in his possession, or to purchase or acquire, any ammunition to which this section applies without holding a firearm certificate in force at the time, or otherwise than as authorised by such a certificate, or in quantities in excess of those so authorised.
>
> Firearms certificates must be readily available for inspection by the police. Failure to do so risks the firearms being seized and detained.

To buy a shotgun in Britain you need to hold a shotgun certificate but if you already hold a firearm certificate you can attach your shotgun certificate to it by paying an additional fee. To buy a rifle you need to hold a firearm certificate.

A firearm is defined as: 'a lethal barrelled weapon of any description from which any shot, bullet or other missile can be discharged'. Air rifles are included within this group if they produce muzzle energy higher than 16.2 joules (12ft/lb), and air pistols higher than 8.12 joules (6ft/lb) at the muzzle. Complete definitions are provided for each of the firearms to avoid confusion. The certificate is valid for five years.

In making your firearms application, you need to contact your local police station, who will put you in touch with a specialist police firearms liaison officer. You need to complete a lengthy application form which contains detailed questions, and requires two 'suitable' referees.

The officer will visit you at home and will need to be satisfied that you have a 'good reason' for wanting a gun, e.g. membership of a clay shooting club, shooting game, or pest control. The officer will also want to see that you will keep your gun(s) and ammunition in a secure place, such as a steel gun cabinet bolted to a wall, and that you have a separate ammunition safe. If you get stuck with any of this you can refer to the Home Office *Firearms Security Handbook 2005*, or take advice from your liaison officer.

Technical Issues and Choice of Firearms

We have included the following information which is intended to be of specific assistance for those who have limited experience in using rifles, and a refresher for others.

A selection of firearms used in pest and predator control.
Left to right: Anschütz .17 HMR, Sako .222 Remington, Browning calibre 20.

The use of different firearms for different quarry is based on a relationship between the muzzle energy of the firearm, the bullet or pellet, and the amount of energy needed to kill the quarry humanely. There is a similar relationship applied to shotgun calibres, load weights, and shot sizes. Muzzle energy is directly proportional to the weight of the projectile and the square of its velocity as it leaves the barrel, divided by a constant. To give an example using imperial measurements, in the case of my 222 Remington, firing a 40-grain bullet with muzzle velocity of 3,445fps (feet per second), the muzzle energy is: 40 (*grains*) × 3,445² (*fps squared*) ÷ 450,240 (*constant*) = 1,054.37ft/lb (or 1429.5 joules).

In other words the muzzle energy is about one hundred times greater than that of a normal air rifle. This information helps us decide which gun to use for each target species.

The Home Office issue guidance documents to police forces and to the public which further illustrate the point. Two of the most useful are: *Air weapons – a brief guide to safety* and *HO-Firearms-Guidance2835[1]*. Both are available on the Home Office website www.homeoffice.gov.uk in pdf format. The chart at the end of Chapter 13 of that document gives guidance on types of gun to be used on various quarry. In the context of this book we are really only interested in the first two quarry categories: *small* e.g. rats or rabbits, and *medium* e.g. foxes.

If you are experienced, you will probably already own at least one weapon. But, if you are a first-time buyer, when it comes to guns for predator control, try to avoid the cost of duplication. For example the HO guidance promotes two firearms, the 17 Remington and 222 Hornet, as suitable for both types of quarry. Both of these use centrefire cartridges (the firing pin strikes a central percussion cap), and both have high muzzle velocities, particularly the lightweight 17 Remington at over 1,220mps (4,000fps). Muzzle velocity is important because it dictates the trajectory of the bullet. A lower velocity (described later), means that the bullet drops quicker and requires more sight compensation at longer distances. Some police forces are now becoming more comfortable with the 17HMR cartridge for foxes, as well as for small quarry. This is a rimfire cartridge (the firing pin strikes the rear edge of the cartridge) and has a reasonable muzzle velocity of 777mps (2,550fps). But beware, the bullet is rather light at 1.1g (17grains) and is therefore susceptible to wind drift at longer ranges.

Our best advice for those using the FAC application approach to predator control is to talk first to your local professional gunsmith. He will have both a technical and a practical view on gun choice, and will also know the attitudes of the local Firearms Department.

UK law allows, by approval, the fitment of moderators. These are sometimes called suppressors or silencers, despite that fact that the result of pulling the trigger is never silence. They moderate the *muzzle blast* and are a major benefit. The subsequent *crack from a supersonic bullet* is still quite loud and cannot be moderated, but it is minimal compared with an unmoderated muzzle blast. The moderator is an added expense but extremely worthwhile.

The guns we use for predator control are:

Falcon T-Bird .22 air rifle FAC 24ft/lb (equivalent to 32.5 joules) for very small vermin.
Anschutz 17HMR for rabbits.
Sako 85 222 Remington for foxes.

Safety

Whilst rigorous, the FAC and shotgun certificate application processes in the UK do not include a formal examination. The firearms officer who visits you will obviously probe the essential areas, and will only grant approval if satisfied that you are not going to be a risk. But, there's no actual test of your competence in respect of handling firearms, safety procedures, or wildlife knowledge. It is therefore essential that you take the time to know your equipment, its capabilities/vulnerabilities and learn the fieldcraft that allows you to engage in a consistently safe process.

Everyone knows that you should never point a weapon – loaded or unloaded – at anything that is not the quarry. On the other hand, if you happen to think the lethal range of a 222 Remington is about 3.2km (2 miles), or that a shotgun is safe because the safety's on, or that the branches of a tree afford a safe backstop, then you need help. Several of the industry organisations offer a range of different services including telephone advice (membership may be required), free website information or training days for all forms of firearm safety. They are well worth the relatively small investment in time and possibly cost, whether you are new to the sport or just needing a refresher.

Equipment Preparation and Practice

Shotguns and rifles should never be regarded as 'take out of the box and shoot' equipment. In the case of rifle/telescopic sight combinations this is a clear and obvious point but equally, shotguns require preparation. A good gunsmith will look at the way you raise and aim a shotgun and may recommend changes to your style and also adjustments to the gun itself. Ideally he will incorporate these adjustments into a 'try gun' that you can take to the range to confirm the adjustments made. After that you should practise before going out to kill things. Shotguns do not have to cope with the distances and fractions of degree accuracy involved in long-range rifle shooting, but the challenges are still there because it is a dynamic process. And, unlike small-calibre rifles, shotgun recoil can have a significant effect upon accuracy. Developing a consistently good swing/mount/point instinct takes some time, but the results can make a huge difference to your accuracy, and overall enjoyment of the sport.

The principles of air rifle preparation and practice are very similar to rimfire or centrefire rifles. The big difference is that the distances involved are shorter. So, for example, where I talk below about zero distances of 90 and 135m (100 and 150 yards), in the case of air rifles, depending upon power, these would be between 27 and 45m (30 and 50 yards).

When selling you a rimfire or centrefire rifle, a good gunsmith will normally offer to set up the rifle/scope combination to a zero distance appropriate to your activities. Incidentally, don't be surprised if his recommendations result in the scope costing as much as, or more than, the rifle: there's no point driving a Lamborghini down the autobahn with a steamed-up windscreen.

In that same context, cleaning kits and good binoculars are essential and an accurate rangefinder is useful. Other items such as slings, bipods/rests and hides will depend upon individual circumstances.

Having got kitted out, and benefited from the gunsmith's assistance, there's still a lot of work to do. With shotguns it is generally a question of consistent technique. The same goes for rifles but, since distance to target and shooting circumstances tend to be more varied, there are a few more things to work on. Our first job with any new rifle/scope is to confirm the '1 minute of arc' (1MOA). This means that, on a calm day in a prone, comfortable position, with a *forestock* rest, we fire three shots at a target placed in front of a secure backstop. A secure backstop means anything that will stop the bullet without risk of ricochet. We always use several planks of wood much larger than the target. This backstop principle must be applied every time you take a shot whether at live prey or artificial targets. It does not mean you have to carry half a tree around with you but it does mean that whatever is immediately behind the target/prey has to be able to stop the bullet safely. In live prey situations a good code of practice is to assume the only safe backstop for a rifle shot is soft ground.

Also note the use of a forestock rest not a barrel rest; you should never rest the barrel of your rifle on anything. Don't even touch the barrel when shooting as it risks compromising the principle of 'free-floating barrels'. This basically means you have to let them do their own thing. The result of the practice shots must be three holes grouped so that, if a circle is drawn around them, its centre will be the bull and its diameter will be no more than 2.5cm (1in) at 90m (100 yards) or 3.8cm (1½in) at 135m (150 yards).

For the initial three shots the most important issue is the diameter (spread). If this is OK and the centre is also on the bull you have a great gunsmith whose shooting style emulates yours. Generally, you will need to make a small adjustment using the windage and elevation turrets on the scope. Having achieved this then, in good conditions, you can expect to be within 1.25cm (½in) of your aim point each time you take a shot from 90m (100 yards) and 1.9cm (¾in) at 135m (150 yards). This sounds demanding but in fact it's not. All of our guns are better than this and one, a Sako 85 with Nightforce scope, consistently creates just one large hole from three shots at 90m (100 yards). A high-quality rifle/scope/cartridge combination is easily capable of 1MOA and, if you don't achieve this, then something needs to change.

Some guns may change their point of impact after a period of bedding in, so don't immediately dash back to the gunsmith after the first three shots. Give yourself and the gun (including cleaning between sessions) a chance to get settled. Equally, make sure your shooting position is comfortable, stable, and consistent. It may save you the embarrassment in rushing back to the gunsmith only to find he takes the gun from you and fires three shots through the same hole.

The Humane Kill

It may sound ironic, but the phrase is used to describe the process of shooting a predator without causing any significant pain or distress. Other control methods like trapping and poisoning may be appropriate, but cannot always provide the same guarantees.

You now have an accurate weapon, and you're technically adept. You go out into the field. A fox comes into range, you take careful aim, pull the trigger and you miss it, or worse still, wound it. Why? I share a passion for both shooting and golf. I do well in both, not because I'm a brilliant shot or a perfect striker of the ball, but because I know I'm not. And, in knowing I'm not, my objective is to eliminate or manage the things that can ruin both sports – the variables.

Setup

The *most accurate* shooting position is *prone*, followed by *sitting* then *kneeling*. *Standing* is the least accurate shooting position. However, often, in the field, the easiest or quickest position to adopt is *standing*, followed by *kneeling*, then *sitting* then *prone*.

I can't achieve 1MOA at 180m (200 yards) standing – not even with a shooting rest – so I don't shoot standing up. I rarely shoot prone, not because of the inconvenience, but because the terrain or absence of a safe backstop can often preclude that position. Although not essential, we have shooting hides positioned near the release pens so most of our shots are taken from a sitting position with a forestock rest and a homemade spring-loaded cantilever midstock rest. Many experienced shooters will consider this to be a totally 'wimpy' approach, but I'm shooting at living creatures, not targets, and I want to eliminate as many of the variables as possible. I want to be in the most comfortable, most stable situation possible before I take the shot.

Everyone has their own technique, but in taking my shot I focus on the amount of movement between the reticle centre and the centre of my aiming point. This movement should be a smooth and equal fluctuation up and down – because I'm still breathing. I then release the safety, take two deep breaths, exhale to the point of comfort, reconfirm aim, and fire. Once fired, I remain in that position for half a second or so. *But*, if something doesn't feel right, I don't take the shot.

Distance to target

If you drop a 222 Remington bullet from head height it hits the ground in 0.693 seconds. If you fire that same bullet horizontally from the same height it also hits the ground in 0.693 seconds. But, during that time, it has travelled almost 0.5km (550 yards). Just before it hits the ground its speed is Mach 1.246 – 1,530kph (950mph). The external forces in play are gravity and air resistance and, fortunately, they are entirely predictable, such that the practical result is that a typical 222 scope/rifle combination, zeroed at 90m (100 yards), needs to be aimed 5.7cm (2¼in) high for a target 180m (200 yards) away.

There are numerous ballistic calculators, many of them free, that work this out for us, including the effect of scope relative to barrel centrelines. The challenge for us is simply to know the distance to the target, and decide on the method of compensation.

We have surveyed all of our hide areas so we know the distance of landmark 'predator hotspots' from the hide. But, being the aforementioned wimp, unless I'm 100 per cent certain, and just in case the predator isn't standing on the box marked 'X', I use my laser-rangefinder/binoculars to confirm the exact distance.

For distance compensation, most shooters use one of two methods:

The elevation turret on the scope. Each click on my Nightforce turret moves the impact point by a quarter MOA which is 1.25cm (½in) (or, in precise terms,

1.33cm/0.5235 in) at 180m (200 yards). So, for the above example, I would turn four clicks up.

The scope reticle. Many different reticle types are available. One example is the mil-dot system which was originally developed for military sniper range estimation. It is possible to use the dots on the reticle as elevation indicators and the accompanying graphic shows the target aim points at various yardages for my 222 Sako.

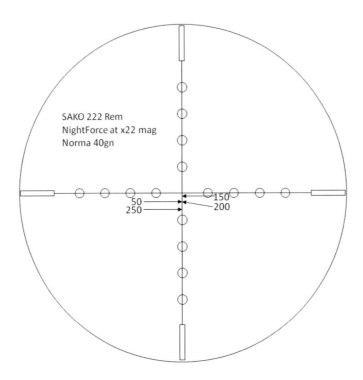

Holdover chart.

Wind effects

Tail and headwinds have a tiny effect upon low-power air rifles but not on anything else. Sidewinds, on the other hand, can be a big issue for all small-calibre firearms. To illustrate this, a typical 222 bullet in a constant 16kph (10mph) sidewind will have drifted 14cm (5½in) from the aim point after 180m (200 yards). Again, ballistics calculators will provide accurate information in this regard, and you can print out windage charts to show the amount of drift.

But determining the actual wind speed and its true direction can be tricky. We have placed windsocks on stakes in our main shooting areas, which give a reasonable indication of conditions, but we don't use them to calculate the offset; they're just there to help us decide whether or not to take a shot.

We're not advising that you can only shoot on a calm day. I've already mentioned that head and tail winds are not an issue. Also, wind generally varies minute by minute in strength and direction, so it normally just requires a little patience to wait for the calmer point between gusts.

Finally, in our experience, high winds also alter animals' behaviour patterns. Some stay in cover and those that don't are generally more nervous and twitchy. In other words, there's always another day, and plenty else to do in the meantime.

The kill zone

Having spent all of this time banging on about accuracy, it might be sensible to mention *where* that accurate shot should be aimed. This may seem blindingly obvious but the margins of error are so small that it's worth a quick revision. There are three target vital organs: the brain, the heart and the lungs.

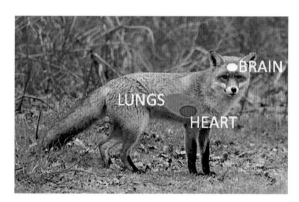

Location of vital organs.

Many people use the term 'head shot' when they (hopefully) mean brain shot. The brain is, of course, a much smaller target than the head. If you shoot at the centre of an animal's head there is a fair chance it could be badly injured but not killed.

A brain shot guarantees death and, normally, no sense of pain. It is the smallest of the three targets and the side-on aim point is midway between eye and ear. Heart shots and lung shots are always fatal but there is likely to be a short period of consciousness. As with humans, the heart and lungs are very close, and present a much bigger target than the brain. Brain shots, even on foxes, are only safe if the animal is reasonably close and not moving, which tends to be a rare event.

There are two important issues in this area of your expertise:

- Knowing the animal's anatomy and *precise* location of its vital organs.
- Using the correct gauge/calibre and cartridge.

Your gunsmith will advise you on the shotgun loads and shot sizes appropriate to your quarry. In the case of rifles he will also be knowledgeable about bullet design. This is important because, unlike rounds used in target competitions, the hunting cartridge you use should be one that, through bullet expansion upon impact, transfers a high degree of its energy to the quarry, and your firearm certificate should have an amendment to permit purchase and use of such *expanding ammunition*.

Shooting at Night

This is often the preferred choice for professional gamekeepers because their quarry species, particularly foxes, are rarely seen during daytime.

Night-time equipment

There are five types of equipment commonly used at night.

Hand-held or vehicle-mounted halogen lamps. Used with standard rifle/scope combinations, ideally this should be regarded as a three-person process. One directs the lamp, a second drives and the third shoots. The advantage is that very high-power lamps are available, which will easily identify quarry out to 180m (200 yards). When foxes are targeted at near distances they will often be spooked. However, unless they have already become lamp-shy, they tend to be curious, but not panicked, at ranges above 90m (100 yards). When used in conjunction with a vehicle the 12V power supply allows virtually unlimited lamping time. The obvious disadvantage is the risk of taking an inaccurate shot, so take care that your setup is correct before you pull the trigger.

Scope-mounted halogen lamps. The most common are lamps mounted on the scope tube and powered from a belt-mounted 6 or 12V battery. Range is more limited with these, as is battery life. I have used these quite effectively for short-range prey but, for convenience and to lessen battery drain, I prefer to spot using a night vision monocular (see below).

Scope-mounted LED lamps. The light source is from high-intensity LEDs and the batteries are usually self-contained. I have an LED Lenser T7 with an effective range of 180m (200 yards) on narrow beam. It gives excellent illumination of the quarry and will easily run all week on a set of rechargeable batteries.

Scope-mounted lasers. These are different from laser-dot sights, and illuminate the target with an intense green light using advanced laser technology. They claim to range out to 550m (500 yards). I have not used one and, whilst this distance is very impressive, 180m (200 yards) at night is the maximum for my fox-shooting skills. And, at ten times the price of my LED Lenser, I would find the cost difficult to justify.

Night-vision (image intensifying equipment). These use image intensifier tubes that gather available light, amplify it, and show the result as a green background image. Animals generally appear as dark shapes. They do not work in complete darkness so they are normally supplied with an infra-red illuminator. The hand-held systems (monoculars and binoculars) used for locating quarry are invaluable aids for night shooting. An adequate monocular can be bought for about £200 at current prices.

Night-vision scopes are expensive, and quite heavy. My husband has an ATN CGTI 4-12x80 Day Night scope which cost just under £3,000 in 2008. The reason for the very high cost compared with spotting systems is not just that it is a telescopic sight. Image intensifier technology is classified in 'generations'. The industry started off with Gen 1 and, in terms of sales to the public, they are currently at Gen 4. The difference in definition is enormous, and my husband would not be happy taking a shot through anything less than Gen 3.

Night safety

For obvious reasons all the rules of daytime shooting apply at night, but in spades. No matter how much has been paid for the best equipment, the scene you will be viewing is not as precise as in daytime, or even dusk. So, take time to develop your observation skills before doing any shooting. With night-vision scopes a cat or a dog can often look like a fox. Observe the type of motion, as well as the apparent shape, of the animal. Foxes are rarely still for more than a few seconds at a time so, if you see a shape that is relatively motionless, it could well be a stalking cat. Ask yourself whether you are absolutely certain about the backstop and, vitally important, *never* shoot at eyes. With any form of night-time illumination the eyes of the animal will present the most vivid image, and that is helpful, but only as a reference to the actual kill zone.

Trapping and Snaring Mammals

Trapping and the Law

Trapping is an effective method of despatching pests and predators, but it must be done properly and, once again, if involved, you are responsible for the outcome. There are many different designs of trap you can use to suit different situations and species; you just need to understand which ones are approved and how and when to use them.

One of the several Acts concerned with trapping is the Wild Mammals Protection Act 1996 which places a duty of care on any keeper of animals, whether temporary or longer term. For example this means you have a *duty of care* to decoy and caught birds, or any species caught in a cage trap. So, for this Act, the focus is not so much concerned with the species or trap, but rather your handling of the trapped creature. An earlier piece of legislation is the Protection of Animals 1911 Act (and its devolved legislation), which still remains relevant today. One of its sections considers the management of traps and states that: 'Where spring traps are used, they should be inspected at reasonable intervals and at least once every day between sunrise and sunset. Failure to do so is an offence.'

Specific legislation for using spring traps is covered under the Spring Traps Approval (England) Order 2012. It revokes the Spring Traps Approval Order 1995 and the variations (England) which followed in 2007, 2009 and 2010 respectively. The legislation is devolved to the other governments in the UK with, for example, the Spring Traps Approval (Scotland) Order 2011 being amongst the most recently amended (24 November 2011).

Under the Acts mentioned it is deemed illegal to use any unapproved spring traps (e.g. a gin trap, which is a form of spring trap with toothed jaws, banned in 1958) for the purposes of killing or taking animals. Approved traps are listed, together with the circumstances in which they can be used and the creatures which they may catch (e.g. a Fenn trap placed on a pole to catch birds of prey would not be approved).

Break-back traps, designed and used to catch small mammals (e.g. rats and mice), or the type commonly used for taking moles in their runs, are allowed, and are dealt with separately under the Small Ground Vermin Traps Order 1958.

It is true to say that legislation is almost constantly changing, so it's vitally important to make sure your understanding is current. However, the best practice guidance is more constant. I have taken the technical information in this section from a number of sources, notably the BASC and the GWCT.

Using Traps for Mammals

Before you use any kind of trap you must conduct a full risk assessment exercise. That is to make sure that the sites you designate for traps conform to legal requirements, and pose no risks to livestock or humans.

Cage traps (live catchers)

Cage traps come in various sizes, depending on the target species. All are basically a box constructed with wire mesh with one or two open ends. The doors are triggered by a foot plate or hook from which bait may be suspended. With the two-door design, you need to set each one independently. Depending on your intended usage, the mechanism works equally effectively when using just one of the doors. When the predator enters the cage, it activates the treadle causing the doors on each end to slam shut, preventing the animal from escaping. They usually come equipped with a carry handle on top of the trap which makes them portable and easy to move around.

Fox caught in a box trap.

We are fans of cage traps. They can be a highly effective method of trapping animals safely thereby allowing an efficient despatch of an otherwise unharmed animal. Owing to their design features, they are also easy to camouflage and bait. Points to note are:

- Cage traps can be used to take any animal normally up to and including fox size which is not protected, e.g. foxes, stoats, weasels, rats, grey squirrels and rabbits.
- You must check the traps at least once every day and despatch target predators quickly and humanely.
- Resist the temptation to simply chuck the body into the woods, or proudly display the carcass on your fence or tree branches. Some gamekeepers may vehemently disagree with this, but there is no scientific evidence that proves this tradition works as a form of deterrent. The most the carcasses are likely to do is put off your average 'right-to-roamer' who might become a little squeamish at the sight. However, it does nothing for the aesthetics of the place, and rotting bodies lying around only encourages disease and further predation. So, dispose of the body responsibly e.g. by incineration or burial.
- It goes without saying that any non-target species must be released unharmed.

We have also used cage traps for completely different purposes. We found, quite by accident, that they can be a highly effective method of catching up stock at the end of the season. We had set a fox-sized trap close to the breeding pens and found two melanistic hen pheasants in it the next morning. We penned them in the 'hospital wing' for a month and put them through a treatment programme to see how they would shape up. They flourished, showing no signs at all of disease, so we put them in with the adult birds. The birds have since become the foundation stock to our very robust group of melanistic pheasants.

Spring traps

Whilst we prefer to use cage traps, there are times when alternative methods of trapping are necessary. Several types of spring trap are on the market and the 'approved' list is constantly under legal review. With a basic design of a trigger plate and jaws, their prime usage is for trapping small mammals. The target list includes rats, mice, grey squirrels, stoats, weasels, mink and rabbits. Different traps have different rules for setting so make sure you understand the specifics of each.

MK6 Solway spring trap. *(Photograph courtesy Solway Feeders)*

Among the wide range of approved spring traps, a popular type used for smaller mammals is the Fenn series. These come in a variety of sizes depending on the type of target animal. To comply with the law, traps need to be set inside either a natural or artificial tunnel/burrow. This is the case with all spring traps with the exception of the Aldrich trap (noted below), which is used for larger carnivorous mammals. It is illegal to set a spring trap on open land.

When you set your trap, try to position it in a predator 'run'. Usually these are fairly easy to recognise and, once established, they tend to be used routinely. Make sure you set the trap with the jaws open to the left and right, in a concealed position. If you can't find a run, look for an appropriate setting site which has thick undergrowth, or along the base of hedges, ditches or a suitable tree hollow. Find a spot to build a simple tunnel. This helps prevent non-target species springing the trap. Typically, we will construct a tunnel using a couple of bits of wood overlaid with turf sods, and liberal sprinklings of leaves with natural debris either side to form a funnel. Any materials will do really, so long as it looks natural and inviting. Once completed, place some kind of natural obstacle like a couple of solid sticks around 7.5cm (3in) apart positioned vertically in front of the trap to avoid trapping the wrong animal, such as one of your own poults or the dog's nose.

According to the GWCT the 'DOC' tunnel trap is another very effective spring trap:

It was designed for the Department of Conservation (DOC) in New Zealand.

Currently only the DOC 150, 200 and 250 traps are available in the UK. The DOC 200 trap has been approved for use to catch grey squirrels, rats, stoats and weasels. The trap must be set in the tunnel provided by the manufacturer for use in the UK. The DOC 150 requires the same tunnel dimensions as the Fenn Mk IV or Springer No 4 traps, an advantage where it must be fitted into a pre-existing space (e.g. in a drystone wall). The two traps have a similar strength, but because of its larger dimensions, the DOC 200 is actually easier to set.

Although legally authorised, the Aldrich spring-activated snare is unlikely to have regular use in Britain. Developed by American hunter Jack Aldrich in the 1960s it was probably intended originally for the control of black bears. It is a pressure-sensitive snare which has a small piece of metal attached to a loop which snares the animal. The metal loop sets the desired circumference. This snare can only be used for killing or taking large, non-indigenous, mammals, which rules out its use in fending off unwelcome in-laws.

Snares

Snares generally are controversial. They can legally be used to catch foxes, rabbits, hare, rats and mink. However, used incorrectly they can cause tremendous suffering in captured and escaping animals, so the key is to use them responsibly. There are several different designs available, which are broadly similar. To describe the main principles I have used below the GWCT Fox Snare design, and have concentrated on fox snaring, which is the most common use of the equipment. I have also referred to the GWCT's 'guidance for the user' information. We would encourage you to read the entire document, which is extremely informative.

(At this point it should be stressed once again that in law *you*, as the operator, are personally liable if something goes wrong with your snaring, so make sure you understand the regulations, which may vary depending on which country your shoot is in. It may help to read either the Defra Code of Practice for setting snares in England and Wales, or Provision 13 of the Wildlife and Natural Environment (WANE) Act (Scotland) Act 2011. The latter incorporates much more stringent rules, including the obligation to undergo training, an ID number, issued by a Chief Constable, for the person setting the snare and the use of ID-bearing tags on snares.)

Design and action. The idea of a snare is to 'hold', not 'strangle', an animal therefore the design is very important.

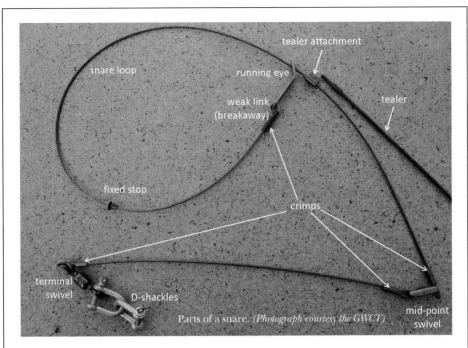

Parts of a snare. *(Photograph courtesy the GWCT)*

The GWCT snare image shows each component part of the snare.

Wire cable: 2mm (½in) diameter cable, with a 208kg (460lb) breaking strain is recommended for foxes. The length should be kept to a minimum to avoid the captured animal getting entangled or running and maiming itself.

Crimps: These join each of the components and can be pressed with a hand tool. The crimp at the noose eye can be designed to break at a specified lower force, which avoids a trapped animal escaping with the noose attached.

Swivel: These are used to work with the animal if it rolls, and prevent it from becoming tangled or weakening the wire.

Eye: This slides along the wire to form the noose. By law it must not be self-locking; conversely it needs enough friction to prevent the snare loop popping open. This isn't always easy but never let the wire become rusty because this may impair its movement.

Stop: This is a very small crimp fitted about 23cm (9in) from the eye that stops the noose closing beyond a predetermined diameter. This prevents death by strangulation and the permanent capture of non-target species.

Tealer: This supports the snare. The best material is copper wire 2–3mm (½–⅛ in) thick. Although expensive it blends in almost anywhere. The attachment to the snare can be made of most materials, but whatever you use make sure it is user-friendly so you can set and un-set the snare easily.

> **Anchor:** This must be robust. The GWCT recommend that a 450 × 25mm (18 × 1in) length of 5mm (⅕in) thick angle-iron driven into the ground will do the trick. An 8mm (⅜in) hole should be drilled 50mm (2in) from the top, in the middle of one side of the angle iron. Connect a 50mm (2in) diameter welded ring to a D-shackle, and fix the shackle to the stake via the drilled hole. Attach the snare to the D-shackle, which acts as a universal joint. *Never* attach a snare to a post or tree. This encourages entanglement and unnecessary injury or death. *Never* use a 'drag' anchor; you may never find the animal that has escaped with the snare wrapped around it.
>
> **Terminal loop:** This is the end of the snare which can be attached to the anchor using a 'D'-shackle.

Setting a snare

Foxes are incredibly wily, as we all know, so you must disguise your snare properly. Make sure it isn't factory fresh. Boil it, rub it with earth, paint it with non-smelly green paint, wax it – anything to get rid of the glossy shine and scent that is unmistakably human. Choose a setting spot with tracks. Look for droppings, bits of fur or that musky smell. Foxes like to walk on easy surfaces if possible. If you have a near miss you'll know because there will be a bit of fur stuck in the wire, but at least you'll know you're in the right area. Use a fresh snare and set it again a little further along the trail – you might just get lucky. If you don't have an instant result, be patient. Foxes do roam and may not visit the same trail every night, so leave a set snare in place for at least a week. The GWCT estimates that in a lowland estate you may need only twenty snares and five nights per fox.

When setting the snare, try not to mess up the run by flattening the area or leaving your scent. Rub soil on your hands and crouch rather than kneel; foxes can detect human scent extremely easily through trousers! The snare loop needs to be set about 18cm (7in) above level ground, but up to 30cm (1ft) high in open ground and around 22cm (8–9in) on banks, with the loop itself measuring 23 × 15cm (9 × 6in). Plant the tealer at a slight incline, to one side of the run. Support the loop (GWCT suggests using florist's wire, which is very malleable), on the opposite side of the run. This keeps the loop nicely and unobtrusively suspended. Alternatively use a longer tealer from which to hang the noose. Try to make the best of the surroundings by bending (or fluffing up) surrounding foliage in the direction of the loop. You're trying to make the snare look as natural as possible. It goes without saying that, when setting your snares, they must not be close to public footpaths or an area where you know dogs or non-target species may be about.

As with the spring traps you should check snares at least twice per day, in summer the first check being before 9 a.m. to ensure that target victims are despatched quickly and efficiently. You also need to have a system for remembering where your traps have been placed. But the main thing to remember is that snares are intended to restrain, not kill, target species so you must use them knowledgeably and responsibly.

Trapping Pest Birds

Legal Issues

If you need to control the predation from corvids you can either shoot them, net them (but using only a net that is propelled by hand, which makes this option fairly impractical), or use a cage trap. So, for the cage trap, let's be clear, under the Wildlife and Countryside Act 1981 section 5 and the Wildlife (Northern Ireland) Order 1985 article 12 it is illegal to:

- Set in position any trap which is calculated to cause bodily injury to any wild bird coming into contact with it.
- Use as a decoy, for the purpose of killing or taking any wild bird, any sound recording or any live bird or other animal whatever which is tethered, or which is secured by means of braces or other similar appliances, or which is blind, maimed or injured.

Section 8 (1) of the Wildlife and Countryside Act 1981 provides that 'if any person keeps or confines any bird whatever in any cage or receptacle which is not sufficient in height, length or breadth to permit the bird to stretch its wings freely, he shall be guilty of an offence and be liable to a special penalty'. Under the same legislation it is also illegal to sell (live) decoy birds.

Traps

Larsen trap
Probably the most popular of its type and named after its Danish inventor is the Larsen cage trap. The BASC describes it in the following way:

It has a closed compartment for confining a live decoy bird and a spring-activated trapdoor which is either top or side entry. The birds legally allowed to be trapped are magpies, crows, jays, jackdaws and rooks. The live traps use a 'decoy' bird, which is kept in one compartment, and when another bird lands on top, it falls through a one-way gate. By law they must have a perch, shelter, food and water.

Larsen trap. *(Photograph courtesy BASC)*

The use of this trap is permitted under a general licence. Failure to comply with the conditions of the licence may render you liable to the penalties already stated.

Multi-catch cages

These can be highly effective if you have large numbers of corvids to control. Once again, under the terms of the appropriate general licence, these can be used to trap the following species: rook, crow, magpie, jackdaw, jay*, great black-backed gull, lesser black-backed gull, herring gull, feral pigeon, collared dove* and wood pigeon.

*Jay and collared dove are protected species in Northern Ireland under the Wildlife (Northern Ireland) Order 1985.

There are several types of multi-catch traps available that conform to three basic design types: the roof funnel, the ground funnel and the ladder letterbox. This last is the most popular for controlling rooks, crows and jackdaws, an added advantage being that it is almost impossible to accidentally trap a pheasant (not sure about a partridge, but we haven't trapped one yet). It is a large timber-framed catcher with the top and sides covered in small wire mesh (maximum 35mm/14in diameter). A ladder-type structure on the top of the pen lets corvids into the trap to feed but they can't fly out.

The BASC recommends the following dimensions: 'Depth 2m, width 2m and length 3m [6ft 6in × 6ft 6in × 10ft] as a minimum size for multi-catch ladder/letterbox type traps. The design should include a door to facilitate daily inspection.

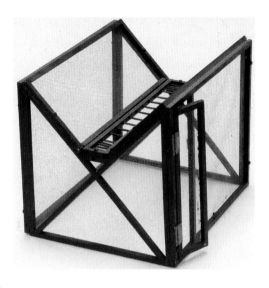

Ladder trap. *(Photograph courtesy BASC)*

Points to Remember

General reminders

1. Use only a permitted bird as a decoy.
2. Immediately release any non-permitted birds that have been accidentally trapped.
3. Provide the decoy bird with a perch, water, shelter and food.
4. Do not wing-clip the bird, there's no need.
5. Legally permitted trapped birds must be killed quickly and humanely.
6. When not in use, make sure the trap is properly closed up,
7. Check traps at least twice daily, either end of the day.
8. Never use a trap, on any land, without permission. This could result in a prosecution.
9. Always remove food remains when the trap is not in use.

Baiting traps and domestic animals

Non-poisonous bait can be used in traps but you need to be careful about where you locate them. If, through being placed near an adjoining field, a trap attracts domestic animals which would not otherwise have been attracted to the land, and those animals are injured or destroyed, then the person setting the trap could be liable to pay compensation and possibly be prosecuted under the Criminal Damage Act 1971.

Domestic cats and dogs can be classed as 'property' under the Theft and Criminal Damage Acts or equivalent legislation in Scotland. Consequently it could be an offence to set traps intentionally or recklessly to kill or injure these or other domestic animals. Obviously we would never consider this an option, but the 'wayward' domestic animal has certainly caused us problems. We occasionally come across a lost dog roaming around causing havoc in the cover crops, and cats lurking around the release pens.

It is extremely frustrating watching your poults get scattered, and damaged in their panic, or hopelessly lost. The problem is worsened by the gundog that realises it's on to a good thing and decides to incorporate your land into its daily routine. At times like this we tend to focus our attentions on the owner rather than animal, which is just having a fun day out.

We have caught up a number of dogs over the years and, if they had an ID-bearing collar, returned them to their owners with a few choice words about responsible pet ownership. This generally lands on 'cloth ears', but does sometimes work. On several occasions footage from our static cameras (discussed in the next section) has helped to demonstrate to the owner in denial that their animal was up to no good, when in fact it should have been tucked away in a kennel. And, in the absence of a collar, we have taken a couple to the vet to establish whether they could be identified through scanning for a microchip or checking a tattoo. If not, I'm afraid they go through the abandoned pet system. The fact is, this is a needless waste of precious time.

Problems with cats can come from one of two sources: domestic or feral individuals. Although, essentially, these are one and the same animal (*Felis catus*), they can be dealt with differently. The family pet is classed as the personal property of its owner. Conveniently for the owner, a cat's owner cannot be held liable for its actions, but could sue for damages should it come to any harm, or go missing. Fortunately, the domestic cats that have appeared on our land have not been too bad to handle. We have generally put them off with a mixture of gusty bawling, and the odd sprinkle of water. This kind of treatment is always effective in dealing with a specimen like our non-mousing, furball of a cat 'Brutus' – *Felis ignavus*!

The behaviour of feral cats is different again and is probably a reflection of their lifestyle. They are prolific killers of nesting birds. If it can be shown that they have bred or are living in the wild then they may be humanely despatched.

Less Conventional Deterrents

This final section summarises other less conventional methods we have tried in addition to the sonic emitters mentioned earlier. Prepared to give most anything a go in an effort to control predation, we have also used the following.

Mechanical/Artificial Devices

Stealth cameras

Obviously best at shooting pictures rather than predators, we have used these as information gatherers to help our deterrent planning. We have four infrared static cameras that are movement-activated. They are set to film for two minutes each 'take'. We position them in what we believe to be predator hotspots. This has often helped us target the correct predator as they are caught on film, which enables us to use the appropriate tool to deal with them. In truth, however, they have been of greater value in capturing some amazing wildlife footage.

Radios

These are a traditional idea we use from time to time. We have a couple of radios fixed on the release pen fences that are switched on during the night. The theory is that the noise will deter foxes and other mammalian predators whilst having no adverse effect on the game birds.

Scarecrows and CDs

Our scarecrows are decorated with aluminium foil strips and freezer boxes. Extras include lengths of striped plastic security tape and CD earrings. Our sci-fi scarecrows are positioned in the release pens and shimmer and flap savagely in the sun and wind. We also hang old CDs in tree branches, which arguably gives rather a Christmassy feel to the pines. They're probably a little disconcerting to your average raptor initially, and especially when it is windy, but we're unsure as to whether they truly terrify the enemy for long.

Decoy owls

We tried 'great horned' owl replicas. They are made of resin, and at 1m (3ft 3in) tall, with glinty eyes and adjustable wings, they look quite fierce on top of a 2m (6ft 6in) high stake. We fixed them in the middle of the cover crops just adjacent to the release pen. Kit like this is meant to deter corvids, who become hysterical thinking that an owl is about to raid their nests, with a helpful side-effect of disruption to any nearby raptors, causing them to seek quarry elsewhere.

At first we were quite pleased with our owls because we did have a significant drop in raptor-related deaths in the release pen. However, having subsequently seen various birds using our owls as perches, day and night, we have to conclude that nature soon acclimatises to static objects and the effect, whilst initially useful, can only be regarded as short term.

Decoy owl.

Crow scarers

These are intended to protect crops that provide food for the game birds, rather than the birds themselves. Crops are inevitably susceptible to wild bird damage, which can, in turn, affect their use for game birds. As you will undoubtedly know, there are many different types of bird scarer. These range from the traditional repeat-fire type, to those that imitate distress calls of herring gull, black-headed gull, crow, starling, pigeon, etc. The downsides are that the culprits get used to them fairly quickly and they can have an unsettling effect on the game birds present! Furthermore, if too noisy, they can become irritating to your neighbours. And finally, if you wanted to help the environment, using loud pyrotechnic sounds does nothing for hedgerow birds' nervous systems. We don't use them, but if you like the idea, try reading the National Farmers Union (NFU) code of conduct on the subject. It is a useful piece and may help you make the right choice.

Natural Deterrents

The home guard

As already explained, we have a number of cockerels – a result of the first set of eggs we were given to incubate. Once we had established groupings with suitable hens, rather than eat the rest we decided to experiment, using them as flock guardians. It doesn't take much detailed observation to spot that a flock of hens will respond immediately to the warning call from a highly protective cockerel. These lads are always on crimson alert, much more so than their pheasant or partridge cousins. One chirrup from him and the girls immediately rush to cover.

As it happens, some of our rearing pens are on the same site as the chicken coops. We noticed that the poults soon learnt to recognise the cockerel's warning call and several either rushed for cover, mimicking the chickens' behaviour, or looked skyward for danger signs. This was a mannerism that was much less pronounced in the pens that did not have chickens close by.

We decided to free-range some suitable (sexually well-behaved and non-aggressive) cockerels in the release pen to see if they would be similarly effective. So far so good. Being the large variety, they appear to be too big for the average raptor. Constantly patrolling the release pen, it genuinely seems that they consider the poults to be part of their greater flock and therefore to be guarded.

In terms of nuisance value to the developing poults the worst that happens

Guardian cockerel.

is that the cockerels consume inordinately huge amounts of their food. So, whilst it seems a little quirky and 'off the wall', we're cautiously optimistic that they may be making some contribution whilst doing no harm to the youngsters or the objective of the shoot.

Encouraged with these early signs of success we are running a further experiment by using a small aviary in the centre of the release pen to hold hens. This causes the lads great interest and ignites their protective instincts. The one obvious downside is that it does nothing for the audio-aesthetics of the shoot. If you are out to impress you may not want to have your shooting chums serenaded by the row coming from a flock of cockerels in the middle of one of your drives. The remedy is simple: if it's too noisy just take them out before the season opens.

Distraction food

This is a technique used by several seasoned gamekeepers. It involves using the carcasses of pests or predators to distract raptors away from poults. Typically these will be rabbits or foxes that are gathered after the kill and put together on a site away from the release pens to become a food source for corvids and raptors. Whilst we could not recommend this practice because of the potential for attracting vermin and disease, the enthusiasts believe it to be an effective device.

Human urine

I'm taking no responsibility at all for the efficacy of this suggestion, but there is a school of thought that believes the spread of human urine around the perimeter fence area is a highly effective deterrent to foxes. It may just have been a personal favourite of the gamekeeper who gave me this tip, in the pub one evening. But just in case…keep the lemonade flowing!

Mix and Match

With many of these simple ideas, we think the key is to mix and match and then change again. Sadly, unlike most of our game birds, the predators are certainly not witless, so you must vary your forms of attack regularly to retain the element of alarm. If you don't, they will learn that the whole thing is a hoax, and ignore it.

So, going full circle really, it all comes back to the overall message of this chapter – pest and predator control is simply a question of good keepering.

9

A FINAL FEW WORDS

'WITH THE BENEFIT OF HINDSIGHT'

Review of Our Objectives

As I explained at the beginning of this book, having bought our property all those years ago, one of our ambitions was to establish a game bird shoot. Nothing fancy, just enough to go shooting with our friends and dogs. For all the reasons I have described it's taken much longer than anticipated, but we have achieved our goals, and are very happy with the result.

There's nothing quite like taking a walk on a frosty winter's morning on your own patch, and watching your dogs at work. You know where to go, because you planted the crops and built the release pens. There's a pretty safe bet that any game flushed is yours, and will provide a challenging, sporting target. You can also be confident that these birds have been nurtured and reared correctly, giving them the best possible start in life. And come springtime, there's the added excitement of waiting to discover whether or not all the effort put into developing a suitable habitat and protecting the game has been sufficient to enable some of them to naturalise and reproduce in the wild. That's what building a sustainable shoot is all about, regardless of the form it takes. It might

Pheasant in flight. *(Courtesy Laurie Campbell)*

be walked-up, driven or rough shooting; the principles are no different. If you love working with animals and you are passionate about the countryside it's genuinely worth all the trials and tribulations.

There is no doubt that we made mistakes along the way. But then again, that is part of the learning process and we have never claimed to be experts. Don't confuse this with being 'gung-ho' though. In a situation where you're dealing with the health and welfare of living creatures, that's out of the question. But things are going really well, and fortunately for the shoot, we tend not to sit on our laurels for long, and each year we try to improve on past performance. There is also no doubt that we have much more to learn. Just fifteen precious minutes spent with a veteran gamekeeper demonstrates that.

Global Reminders

Since the purpose of writing this book is to offer early guidance on a number of topics where either scant or confusing advice often exists, it would be remiss to end without some kind of global summary. So here are some final reminders:

- The shoot took much longer to develop than we had envisaged. Regardless of the number of birds you plan to release, you will not produce a successful shoot in six months. The methods you use to rear your birds and the point at which you begin your hands-on involvement

The Professionals' Postscript

Throughout this book I have quoted and referenced several recognised experts who have helped us in each subject area. Much of this information is freely available from the many organisations that represent our industry. Still more is accessible through membership. We believe that an affiliation to one or more of the 'best in class' associations/trusts is essential. You become part of the gamekeeping community, and benefit from 'on-tap' non-judgemental practical advice.

Your choice of affiliation is wide, and to point you in the right direction I asked each of the key organisations that I have consulted or cited to add a comment about their business and also to offer a word of advice to new gamekeepers. Their responses follow.

The British Association of Shooting & Conservation (BASC)

BASC exists to promote and protect sporting shooting and the well-being of the countryside throughout the United Kingdom and overseas. There's plenty of information on the website which is free to anybody and, as a member, you can get immediate personal advice by phone or email from expert staff on all aspects of shooting, gun ownership and game management. Other benefits include full insurance cover, trade discounts, training courses, Young Shot events and the UK's biggest circulation shooting magazine free. BASC is the voice of shooting at Westminster and maintains a strong presence in the devolved administrations and in Europe.

Glynn Evans, Head of Gamekeeping for BASC, says:

> To be a good keeper, whether full-time or amateur, requires many different skills. It is vital that you understand the complex strands of game and wildlife management and how they interact with each other. But it is certainly within the capabilities of most people to achieve this by learning as they go along. The key to this is 'never be afraid to ask'. Experienced keepers will tell you that they themselves are often seeing new things and learning new ways.
>
> So search out information and new ideas. Books, magazines and the internet are all worth a look and do not forget the value of talking to others and drawing on their experience. We can all learn from mistakes, but it is easier and less costly to avoid them in the first place.
>
> Remember to plan ahead. Understand and comply with legislation, best practices, and never try to cut corners. By maintaining the highest standards success and good results will be achieved.

The Countryside Alliance Foundation (CAF)

David Taylor, Shooting Campaign Manager of the Countryside Alliance, says:

> An amateur keeper would find many benefits from joining the Countryside Alliance. As a member you are supporting an organisation that promotes and protects the sport of shooting. You are also part of a community which recognises and celebrates the rural way of life and defends all those who live or work in the countryside.
>
> For an amateur gamekeeper just starting out, I would suggest learning as much as possible about birds, their habitats and the countryside in general. There is no one right way of doing anything when it comes to managing wildlife, but there are many wrong ways. Spending time observing the countryside and its interactions is one of the best ways to learn, but gaining experience from a knowledgeable gamekeeper is second to none. Mistakes through inexperience can be costly in monetary and environmental terms, so never be afraid to ask questions. Take things slowly, and build your knowledge and confidence over time.

Game Farmers' Association (GFA)

Charles Nodder, who writes for the sporting press and is adviser to the GFA and the NGO, says:

> The GFA is mainly for those who rear and sell significant numbers of game birds or their eggs and is the representative body for game farmers. As a trade association, membership is more expensive but, like the NGO, the GFA produces some excellent and informative publications, including a large ring-bound GFA Game Farming Guide that members get free and which is regularly updated. There is a category of Associate Membership for those who are not game farmers as such.
>
> The main thing for a newcomer to the industry is to be a member of something. It's an opportunity to make a contribution and be represented. Each of the top organisations offers good value although the precise membership packages will vary. Support as many as you can.

Game & Wildlife Conservation Trust (GWCT)

Dr Mike Swan, Head of Education with the GWCT, says:

> The GWCT has an unrivalled breadth and depth of knowledge in conservation and game management. Being a member helps to keep you abreast of this, and

eggs
    anatomy of *42*, 43
    artificial clutches 45, 124, 131
    buying in 34, 58
    collecting and handling 45, 52–3, 54, 127
    embryo weakness 57
    fertility 34, 56, 57, 59, 60
    storage 34, 58–9
    sub-standard eggs 54–5, 56, 57–8
    surplus eggs 126–7, 131–3, 213
    washing 54, 59, 142
    see also hatch; incubation
electric fencing 100, 101, 115
Environmental Stewardship Scheme 161–2

feather-pecking 46, 76, 84–5
feed rides 102, 105, 116, 213
feeders 41, *41*, 80, 84, 91, 105, *116*
    auto-feeders 91
    cleaning 84, 92
fencing 88–9, *88*, *89*, 99–100, *99*, 115, 116
firearms
    legal issues 183–6
    see also shooting for pest and predator control
flea-beetle 152, 157
food
    additives 82, 98
    anti-parasite treatment 82–3
    broody hens, feeding 124
    immune system, boosting 139
    maggots 74–5
    pellets 82
    photodegradation 75
    pre-release 112
    second age granules 82
    starter crumbs 74, 82
    titbits 39
    see also feeders
fount drinkers 79
foxes 181, 182, 191, *191*, 192, *195*, 198, 200
French partridge *see* red-leg partridge

Galliformes 16
Game & Wildlife Conservation Trust (GWCT) 97–8, 114, 115,
129–30, 174–6, 198, 200, 215–16
game crops 115, 118, 144–62, 213
    bird scarers 207
    catch crops 152
    climatic variations 146–7
    conservation practices 159–60, 161–2
    crop mixtures 151–2, *151*, 155, *160*
    crop purposes 145, 153, 154
    grant schemes 161–2
    kale 148, 149, *149*, 156, 157
    maize 147–8
    millet 150
    organic crops 152–3
    perennials 150
    pest control 156–7, 159
    preparatory work 153, 154–5, 156, 159
    problems and solutions 157–8
    quinoa 150, 156
    rotation 154
    seed quality 159
    soil analysis 145–6
    sorghum 148–9, *148*
    weed control 154–9, 159, 213
Game Farmers' Association (GFA) 13, 32, 34, 215
game farms 32–3, 34, 36
game meat cookery 132–3
Game-to-Eat initiative 132
gamekeeping organisations 214–16
gamekeeping profession 214–17
gapeworm 136–7, *137*
'gentle release' system 94–7
gestation periods 66
gin traps 194
golden pheasant 19, *19*, 21
grey partridge 26–8, *28*
grit 38, 40, 91

hardening off 70, 77, 83, 108, 212
hatch
    assisted hatching 55, 62–6, *64*, 65
    hatch dates 61, 66–7, *66*
    hatch rates 34, 55–6, *56*
    hatch states
        death in the egg 51, 62
        peeking 61, *62*
        pipping 61, 62, 125
        wrong enders 61
        zipping 55, 61, *63*, 64
    hatching incubator 49, 50, 70
    premature or overdue chicks 66–7
heaters 81
hen coops 122–4, *123*
hospital pens 135
Hunting Act 2004 (England and Wales) 181
hybrid turkens 121, *121*
hygiene 38, 39, 84, 134–5, 142, 143, 212
hygrometers 49, *49*, 52

immune system, boosting 138, 139, 140, 142
incubation
    artificial 12, 44, 46–68, 131
    natural 25, 30, 44, 52, 127–9, 131
    process
        candling 59–60, *60*
        date-labelling the eggs 53, 54, 127, 128
        egg selection 53, 54–8
        gathering eggs 52–3, 54
        setting the eggs 53–4, 59
    see also broody hens; hatch; incubators
incubation suite 70–6
    hatching incubator 49, 50, 70
    post-natal nursery 70–6, *73*
incubators 46–52
    alternative uses for 67–8
    automatic rotation 51, 66, *66*
    basic design 50–1
    buying 50–1
    cleaning 134–5
    forced air systems 47, 48
    hatching incubator 49, 50, 70
    humidity 47, 48, 49, 50, 51, 52
    manual rotation 51, 54, *66*
    number of 49
    self-build projects 48
    still air systems 47–8, *47*, 50
    temperatures 47, 48, 49–50, 52
infra-red heaters 81

INDEX

kale 148, 149, *149*, 156, 157

ladder traps 203, *203*
Lady Amherst's pheasant 20, *20*, 21
lamping 192–3
Larsen traps 201–2, *202*
laying
   cycles
     partridge 44, 45
     pheasants 44–5
   nesting sites 45, 127, 128, 130
   raised system 34, 46
   random egg laying 45
   signs of 45
lead shot, use of 170–1
leg malformations 71–2, 73
   crinkle toe 71
   splay leg 71
   splinting technique 72, *72*
legal issues 162–71, 213
   'access land' 170
   accident insurance cover 168–9
   firearms 183–6
   lead shot, use of 170–1
   pest and predator control 174, 177–81, 194, 198, 201
   public access rights 169–70
   risk assessment and management 162–7, 169
   shooting leases 168
   trapping and snaring 194, 198, 201
   trespass 170
licences for predator control 179–81

maggots 74–5
maize 147–8
medications 139–40, 142, 143
mice and rats 172–3, 182
millet 150
misting 83–4

National Gamekeepers' Organisation (NGO) 46, 216
Natural England 161, 162
nesting pens *see* breeding pens
netting 89, 99, 114
night-shooting 192–3
night-vision scopes 193

nipple drinkers 35, 78–9
nurseries
   brooders 70, 76–83
   construction 76–7
   feeders and drinkers 78–80, *79*
   floor coverings 78
   hardening off process 70, 77, 83, 212
   heating 80–1
   height 78
   playthings 81
   stocking density 76
   temperatures 77, 83
   cleaning 84
   post-natal nursery cages 70–6, *73*
     base covering 72–3
     escapees 76
     heating 75
     wire base 74
   runs 70, 83–4

organic crops 152–3
overcrowding 139, 142

parasites 32, 33, 38, 39, 40, 141, 142, 172
   coccidiosis 82, 141
   gapeworm 136–7, *137*
   rodent carriers 174
   symptoms 141
   treatment 82–3, 141
partridge 26–31
   grey partridge 26–8, *28*
     characteristics and habits 28
     chicks 27, 28
     description 27
     eggs 28
     hens 27, 28
     native range 26, 27
   red-leg partridge 26, 28, 29–31, *30*, 33
     characteristics and habits 29–30
     chicks 31
     cocks 30
     description 29
     eggs 30, 53, 55
     gestation period 66
     hens 30
   species and sub-species 26–31
pens
   breeding pens 34, 129–30, *130*
   construction 37–8

existing facilities, using 37, 212
hen coops 122–4, *123*
hospital pens 135
raised pens 46
   see also nurseries; rearing pens; release pens
pest and predator control 172–209, 213
   control options 182
   corvids 201–5
   effects of good keepering 174–6, *175*
   game crops 156–7
   hunting with dogs 181
   legal issues 174, 177–81, 194, 198, 201
   licences 179–81
   mechanical/artificial devices 89, 205–7
     crow scarers 207
     decoy owls 206, *206*
     radios 205
     scarecrows and CDs 205
     stealth cameras 205
   natural deterrents 207–9
     distraction food 208–9
     guardian cockerels 67–8, 102, *103*, 104, 207, *208*
     human urine 209
   parasites *see* parasites
   poisons 173–4, 183
   rodents 172–4
   shooting *see* shooting for pest and predator control
   trapping and snaring *see* snares; traps
Phasianidae 16
pheasant
   common pheasant 23–6
     characteristics and habits 24–6
     chicks 25–6
     cocks 23, *23*, 25, 26
     description 23–4
     eggs 25, 53, 55, 58
     gestation period 66
     hens 23, 24, *24*, 25
   history 16–18
   native range 16–17
   species and sub-species 16, 18–21
photodegradation 75
playthings 81, 84, 212
poisons for predator control 173–4, 183

post-natal nursery cages 70–6, *73*
poults
    buying in 35–7, 106, 107–8
    catching 35, 111, 117
    dogging-in 118–19
    transporting 36, 38
    wing-clipping 35, 108–9, 110–11, *111*, 117
        see also rearing pens; release pens; releasing poults
predators *see* pest and predator control
premature or overdue chicks 66–7
Protection of Animals Act 1911 194
Protection of Wild Mammals (Scotland) Act 2002 181
public access rights 169–70

quinoa 150, 156

rearing pens 85–92
    construction 86–9, *86*, *88*, *89*
    'fallow farming' system 91
    feeders and drinkers 91
    furnishing 89–91, *90*
    hides and shelters 89–90
    maintenance 92
    routine husbandry 91–2
    stocking density 86
    transfer process 91, 213
rearing process
    choosing which bird to rear 15, 18, 212
    getting started 31–2
    overview 211–13
    sourcing stock 32–3
    timeline 14, *14*, *70*, *126*, 211–12
        see also buying options
red-leg partridge 26, 28, 29–31, *30*
Reeves' pheasant 21–2, *21*
release pens 93–119
    call birds 104
    construction 99–104, *99*
    feeders and drinkers 105, 116
    fencing 99–100, *99*
    'gentle release' system 94–7
    guardian cockerels 102, 104, 207–8, *208*
    GWCT guidance 97–8, 115
    maintenance 112
    open-top release system 97–8
    partridge pens 113, *113*, 114–16, *115*, *116*
    pop-holes 97, 99, 100–1, *100*, *101*, *102*, 108
    routine husbandry 112
    shelters 105, *105*, 115
    siting 94, 98–9, 115
    transfer process 105, 110–12, 117–18, 213
releasing poults 106–8
    at six weeks 106–7
    at eight-plus weeks 107
    batch releasing 109–10, 117
    partridge release process 98, 113–18
    pheasant release process 94–112
rental agreements 144–5, 168, 171
risk assessment and management 162–7, 169
roaming instinct 118
Rock partridge 29
rodenticides 173–4, 183
rodents 172–4, 182
routine husbandry 38–40, 84, 91–2, 112
    see also hygiene
runs
    hen coops 123
    mini-runs 83
    nurseries 70, 83–4

scarecrows 205
scope reticles 190, *190*
shoot management 163–71, 216–17
shooting for pest and predator control 183–93
    choice of firearms *174*, 185–6
    distance compensation 189–90
    firearms law 183–6
    humane kill 188–93
    kill zone 191–2, *191*
    licences 183–4
    moderators 185–6
    night-shooting 192–3
    preparation and practice 186–8
    safety 186, 193
    wind effects 190–1
shooting leases 168
shotgun certificates 183
snares 198–201, *199*
    legal issues 198
    setting 200–1
sonic emitters 173
sorghum 148–9, *148*
sourcing stock 32–3
    see also buying options
splay leg 71
spring traps 194, 196–8, *197*
stale seedbed technique 156
starter crumbs 74, 82
stealth cameras 205
stocking densities 76, 138, 139, 142
stress 35, 36, 84, 85, 139, 142
survival skills 110
syndicate shoots 171, 216–17

temperament 39
    see also aggressive behaviour
traps
    baiting 133, 204
    break-back traps 194
    cage traps 195–6, *195*, 201–3, *202*, *203*
    decoy birds 202, 203
    double-door box traps 117
    ladder traps 203, *203*
    Larsen traps 201–2, *202*
    legal issues 194, 201
    multi-catch cages 202–3, *203*
    spring traps 194, 196–8, *197*
    see also snares
trespass 170

unhealthy birds 32–3, 135
    see also disease

Verm-X 83, 109
veterinary care 137, 212
visitors, rules for 39

weasels 182
weed control 154–6, 159, 213
welfare issues 12–13, 46
wet weather conditions 40–1, 90, 143
Wild Mammals Protection Act 1996 194
Wildlife and Countryside Act 1981 15, 177–9, 201
wing-clipping 35, 108–9, 110–11, *111*, 117
worming 82, 109